FARMI

1928

BY

W. M. TOD, M.A.

CAMBRIDGE UNIVERSITY AGRICULTURAL DEPARTMENT

WITH ILLUSTRATIONS BY

LUCY KEMP-WELCH

THE PROMISE OF THE YEAR.

CONTENTS

LIST OF ILLUSTRATIONS

FARMING

CHAPTER I

The Best Pursuit of all

OF all the delightful pursuits that one can be engaged in there is none, I am of opinion, that can compare with farming. Dean Hole, in his charming book on "Our Gardens," would have us believe, nay, would almost convince us, that gardening was the queen of pleasures, but it is the farm, that extended garden, that reaches the acme of rural delight. That farming is a business and requires a considerable amount of capital and no mean amount of administrative capability, perhaps removes it for many from the list of pleasures. Alas that hard times of low prices should ever have robbed the land of so many of its pleasures, should have lengthened the jovial faces of our old-time farmers,

A

and hardened the lines round the mouth of the kindly squire.

It is indeed a matter of some difficulty nowadays to make farming pay; but, with the requisite knowledge, skill, and business powers, it is still possible to get a better interest on one's capital than such investments as consols afford, and (with a very big "and") to enjoy such health and strength as a king might envy. To him who farms in earnest, whether for a living or for pleasure, there comes no *ennui;* he never possesses a liver, or lungs, a heart, or nerves; his snack of bread and cheese and beer is a meal fit for a prince, his roast-beef is the best that old England produces, and his sleep the sleep of the just. While about his farm, his eye is full of the beauties of Nature; he watches his carefully tilled soil to see the seed germinate and the tender plant push its head into the light, he sees the grass luxuriate in the spring-time sunshine and shower, he welcomes the cuckoo and the swallow as the harbingers of spring; he watches with a pleasure almost of love, the opening of the hawthorn blooms on the hedges, the ever-changing green of his fields, the blowing of the wild roses and the ears peeping out of the wheat and oats, the squawking, scuttering partridges with their new-hatched broods, the ripening gold of his grain and the purple tresses appearing on the bramble; aye, and a thousand other delights appeal to his eye, his ear, and his nose.

I know of no keener delight than that with which I hasten home on a Saturday night, after a week in town, to spend one day at least in my fields. Directly after morning church and an early lunch, I begin my walk and prowl the fields till darkness drives me in. How everything has altered in that one short week. Here is a field newly ploughed, there a field which I saw last in straight furrows, now finely harrowed down for sowing, and we test the depth of its tilth and judge its suitableness for the seed. A little farther on is a field all scored with the pale yellowish green lines of the newly-shooting barley, and close by is another in whose soil we scratch to find the grain which last week we saw hard and dry, but now has tiny tufts of hairy rootlets.

Yonder is the shepherd ready to greet me with a smiling face, and so I know that all is well and he has new charges for me to see. His lambs are proudly counted and their points admired; we know each ewe and all her history. No frightened looks have these sheep; they rather cluster round and draw attention by their plaintive baa. We must see one sick sheep and advise as to remedies, and after discussing the supply of roots and grass I make my way to the stockyards. How these fattening beasts have altered, to be sure; though I dare say no one but myself would notice it. The foreman evidently notes my pleased smile, and feels that he has done his duty. Each creature has his points discussed and wriggles to and fro while I scratch the top of his tail. They know us as well as we know them, and have never learnt to connect a man with anything but scratchings and a food-basket. The stables are next visited, and the condition of each horse is noted with a pat or two, and a " Hullo, Dobbin," or " Well, old Depper," for each one, which caresses I must say they receive with less display of reciprocity than the bullocks. But the time is getting on, and there are several more fields to look at, and so by the time my walk is over it is very nearly dark, and the arm-chair by the fireside has strong allurements.

Even when I am at home all the time these pleasures

never pall, for Nature is ever changing, and then there
are pleasures of a baser sort, when we see the luxuriant
hay being built into mountainous stacks, or piles of
golden grain carried into the barn from the threshing-
machine; then we feel that this is the reward of our
labours, the surplus over and above the pleasures we
have already received.

Besides these many pleasures, farming has also its
trials, which seem to be sent to teach us at the same
time both self-control and self-reliance. The fretting,
fuming man who gets in a stew, and works himself up
into a state of violent excitement because a mare is
rolling about with cholic, or explodes like a miniature
Krakatoa when a lad chops up a row of swedes instead
of the thistles, or who consigns the weather to perdition;
this man will never make a happy jovial farmer nor taste
life's fullest pleasures.

Then there is the grumbler, improvident and un-
thinking; he never sees the beauties of Nature nor
rejoices in the glorious works of the Creator, and life
is to him a struggle and nothing more. I know him
well and meet him often. He grumbles at his men and
consequently gets all the ne'er-do-weels of the village
in his employ; he grumbles at his landlord and his rent,
and his pet aversion is the village school; it always
rains when he wants to carry his hay or is dry when he
sows his turnips; nothing is ever right, and his life is a
life of misery.

These trials—for trials they are—give plenty of scope
for the practice of self-control and self-reliance, and the
man who thoroughly understands his business and
makes his plans beforehand can nearly always manage
to meet them half-way by being prepared. When it
pours with rain at the time we should be sowing our

wheat, we can limewash the insides of the hen-roosts and cow-houses, or clean up the rickyard ready for threshing; and if our drains are in order and the ditches cleaned out, it will only be a day or two before the fields are dry enough to work again. In frosty weather when the ploughs stand idle, there is dung to be carted, stones to be put on the roads, hedges to be trimmed, harness to be cleaned and mended, and any number of other things which we can tell ourselves would never have been done but for the spell of frost. There are numbers of little jobs which should never be undertaken during fine weather, but carefully saved for wet and frosty days. I well remember a farm bailiff whose chief delight seemed to be to make compost heaps and cart stones when it was fine, and to plough when the water ran in front of the plough. The consequence was he used to sow his oats in June and lost his employer many hundreds of pounds.

A little foresight and skill will usually allow us so to prepare our ground that the barley can be drilled between the showers, or to make sure of a crop of roots in the driest of seasons; and even in that most provoking of all weathers, a showery hay-time, the root crops will benefit by the attention we are able to give them in the intervals.

I know that at times animals will be ill, and the death of a cow or a horse will make serious inroads into our profits, but I have nearly always found it possible to trace the illness to some neglect, and over and over again have found that the immediate calling in of the veterinary surgeon has resulted in a shorter bill and the recovery of the animal.

Much of the neglect of precautionary measures is the result of a legacy from the old times when wheat made from sixty to seventy shillings, and many small losses were looked upon with indifference, as everything would be made right at harvest. The jolly old boys, the portraits of whom one so often has seen in *Punch* and elsewhere, associated with every kind of sport, had merry and plenteous times in which to live. An acre of wheat realised more than double it does now, beef and mutton were practically the same price, labour cost very little more than half the present rate, and his rent and taxes together were but a little if any more than now. His pigs with their three or four inches of solid fat, his dairy and his garden supplied him with the bulk of his needs, and his wife and daughters toiled far harder than he. I have known several of them, and have enjoyed their hospitality many a time, when the table groaned beneath its load of the good wholesome fare that has made the typical Englishman what he is, and where the great jug of good old home-brewed ale has been replenished more than once. Aye, and what a sportsman he was, a sportsman in its truest sense, whether with greyhound, horse, or gun.

The horse was always his special delight, and you can still see him, even if the times have altered sadly, out with the Quorn or Pytchley and a few other Midland packs, although he is lost entirely to many parts of the country.

Farming always bred sportsmen, and I sincerely hope it ever will. The modern farmer, though he has to stick more closely to business, and has replaced the sickle with the self-binder and talks of phosphates and nitrogen rather than of trotting nags, is still a sportsman at heart. Though he can no longer keep his hunter as did his father, he can, when the hounds are meeting within two or three miles of his house, still feel the tingle of his sporting instincts, as he burnishes his stirrups and bit, or grooms down the mare that draws his trap to market, preparatory to joining in the meet. With what excitement again, he cleans his gun and counts his cartridges, when he has been asked to join a party in the squire's coverts to finish off the rabbits and thin out the cocks. Would that landlords more often considered their tenants in these matters, for they could many a time turn a young poacher into a true and keen sportsman. No landowner can have a keener keeper than the young farmer who knows he will be asked to shoot two or three days during the season, and woe betide the cat or stoat that shows itself on this farm.

Sport and farming have ever been closely related, but the hardening of the struggle for existence has unfortunately done much to alienate them. In these days when every bushel of wheat or barley and every truss of straw or hay is a matter of serious consequence, no farmer can be expected to view the tracks of hunters across his wheat with indifference, or to smile when the hares cut paths through his oats, and the young rabbits clean them off altogether by the side of the covert. The competition of the foreigner, who floods our markets with his surplus products, and the competition between shipowners in the cutting of freights, has had much to do with the changed conditions. Farming is becoming a business where the keenest attention to details is required, where the utilising of every yard of ground and the greatest intensity of culture becomes important. The margin for profit on each quarter of grain has become so much smaller that it is now imperative to grow more quarters on the same area, and to secure those quarters with the least possible expenditure and with the minimum of waste. Even although beef and

mutton realise quite as good prices as they did thirty
or forty years ago, the greatest skill and care is required
to produce them at a profit, for we can no longer afford
to use cakes and meals with a lavish hand.

In the days when grain was the chief source of profit,
the farmer who could make his beasts and sheep pay for
their cake considered he had done well, for the fertilising residue left upon the land was sure to increase the
profitable bushels, and was regarded as a sufficient
return for the hay, straw, and roots consumed by the
animals. At that time, every extra bushel of wheat
produced meant seven or eight shillings; but now that
it is a matter only of from three to four shillings, many
modern farmers are finding that the game is not worth
the candle, and are carefully reducing their cake bills
to the lowest possible amount. I shall discuss the
matter of bought feeding-stuffs more fully later on, but
I should like to say here that I regard purchased foods
in the same light as I regard a savage watch-dog, useful
in its place, but to be kept carefully chained up, or
mischief will ensue. I find more and more practical
feeders among those who keep a strict account of their
transactions, who are bearing me out in my opinion
that much of the loss many farmers suffer is due to
the reckless and injudicious use of purchased feeding-
stuffs. This reckless use is as a rule most glaring on
landowners' home farms, but is responsible in many
other cases for the difficulty in successfully meeting
foreign competition.

In stating my opinion thus emphatically, I have no
desire to array myself amongst the farmer's critics, for
I know full well that the many conditions of farming
give rise to many different practices, each of which

may be right in its own particular place. This is often forgotten by the critic and would-be adviser, and no race of men have so many critics and advisers as farmers.

Just as every one considers himself a judge of a horse, so everybody seems to think he can advise the farmer as to his business, and criticise his methods. I have seen some of the most extraordinary advice given in letters to the newspapers. I remember when we had a wet hay season a few years ago, reading a letter in a daily paper wherein the writer pointed out to farmers the method of drying hay adopted by the Swiss peasants, who hung their grass over erections of sticks so that it was kept off the ground and the wind could blow through it, and advising the British farmer to do the same. This might be a very excellent plan for a field the size of a tennis lawn, but it would be absolutely impracticable for even five acres, to say nothing of from one to two hundred acres, such as I and many others have to secure.

Then again I have seen both spurry and groundsel recommended to farmers as forage crops, for what reason no one could imagine, when we have dozens of plants far more suitable, and not likely to cover our own and our neighbour's fields with weeds. The people who recommend these kind of things do so out of the fulness of their ignorance, and may be laughingly excused; but there is another class of person that we may call the faddist, who, having a certain amount of reason and experience on his side, nevertheless fails to see that his particular fad is not suitable for all cases and conditions. The kind of man I mean is the one who runs riot over *Petite Culture*, or poultry and bee keeping, or who makes out that the salvation of farming lies in deep cultivation, deep drainage, or in silos or lucerne growing.   Now we must beware of condemning entirely these or many other things because some one rides his particular hobby to death, for they are all useful in their place, and ours may be at some time a place where one of them may be found useful.   Where the faddist usually makes a mistake is in experimenting with a small quantity, and then, because his results are successful for this small quantity, proceeding to multiply his figures by ten or a hundred as a proof of its profitableness for large quantities.   A tenth of an acre, well tilled, may yield a quarter of wheat, but one acre treated in exactly the same manner will not yield ten quarters ; or twenty fowls may show a profit of five pounds for a year, but experience shows that two hundred fowls will not give a profit of fifty pounds.

Petite Culture may be excellent where the man with his wife and children do all the work, and poultry and bees make first-class additions for bringing in a few pounds, but, as soon as labour has to be paid for, the profit disappears.

Deep cultivation on the heavier soils, and the deep drainage of wet, porous lands, are both excellent in their places, but the crop returns are seldom large enough on a farm to pay for the extensive outlays of capital which have been sometimes recommended.

Silage is an important aid to the stock-keeper, and at any rate a small patch of lucerne should be found on every farm ; but in none of these things alone is to be found the salvation of agriculture.

The enterprising and thoughtful farmer considers how far any of the suggestions made by others are suitable to his land, conditions, and pocket, and tests them in a cautious manner.

It has always, however, been the man of scientific mind who has brought about improvements in farming practice; the well-read farmer who has realised the possibilities, and has sought the aid of science. Up to the end of the seventeenth century agriculture in this country had remained dormant, it is doubtful, indeed, if it was equal to that of the best practice of the Romans. In the eighteenth century a few enlightened men seized upon the crude teachings of science, and with the introduction of the turnip, these few men began to drag British farming out of the mire. Unfortunately the sciences were themselves in so backward a condition that they had very little of value to teach. Manuring had advanced nothing since the time of the ancients, for bones, horn, and wool had long been known, and even nitre was known by the Romans to have an effect on plants.

Scientific research was, however, rapidly advancing, and it became possible, in 1812, for Sir Humphry Davy to deliver his eight epoch-making lectures on agricultural chemistry before the Board of Agriculture—lectures which drew the attention of the world to the connection of chemistry with farming, and laid the foundations of agricultural science.

Progress was now rapid, for many famous scientists devoted themselves to the investigation of the constituents of plants and of soils, and discovered the connection between them. Liebig enunciated his famous mineral theory as to the food of plants, which, although afterwards proved to be incorrect, laid the foundation of the whole of the important and far-reaching subject of the use of artificial manures.

In 1834 Lawes began experimenting with plants, and having discovered the value of mineral phosphates dissolved in acid, he patented a process for the manufacture of superphosphate in 1842. A year later, having secured the services of Gilbert, he started his wonderful experiments at Rothamsted. Botanists had been doing their share in this work, and had gained a considerable knowledge of how plants feed and grow, and in what forms they require the food they use. The uses of phosphates, salts of potash, and nitrogen as aids to plant growth were becoming known.

A demand having sprung up for these fertilising substances, the supply of animal and vegetable manures in the forms of dung, guano, and bones, ceased to be sufficient for our requirements. Lawes had already taught the manufacturer how to convert mineral phosphates into superphosphate; others showed him how to make sulphate of ammonia from the refuse of the gas-works, how to extract nitrate of soda from the "caliche" found on the plains of Chili, the great value of the potash salts dug from the mines of Germany, and quite recently, the properties of the slag (basic slag) which is formed in the crucibles used for the manufacture of steel. In such practical ways as these has science assisted agriculture.

We have become so accustomed to the services of qualified analysts in safeguarding our interests in the matter of cakes and manures that we are apt to forget that we owe all the methods by which this work is done to the painstaking investigations of the earlier chemists. The late Dr. Voelcker deserves mention for his work in the interests of British farmers, though many others have added and are adding their contributions to our

knowledge of these matters.   It is impossible to mention even the names of the very many workers who
have aided in this wonderful work.   I can only point
to the results ; results which have removed agriculture
from a rule-of-thumb art into the realms of a science, and
made its practice possible in these days of low prices.

The long and painstaking investigations of many
physiologists into the digestive processes of animals,
and the consequent value of the various food-stuffs, have
resulted in the placing of our rules of feeding upon a
sound basis, and have shown us how to produce meat
or milk with the greatest economy.

In consequence of the work of these scientists the
old-fashioned cow-doctor has been transformed into the
veterinary expert, by whose teaching and aid our farm
animals are maintained in a state of health previously
unknown, which has resulted in infinite saving to
farmers.

The discovery of bacteria by Pasteur, and the subsequent work of hundreds of investigators, have had much
to do with this result, for the discovery of the cause of
disease has led to the means of its cure.   In the dairy
the work of the bacteriologist has been of infinite value :
it has shown us the causes of failure and the way
to success ; has made it possible to produce butter
and cheese of first-class quality with the greatest
certainty, and has demonstrated that we can produce
any kind or any flavour at will, quite independently of
the locality.

The important bearing of bacteria on the fertility of
the soil is gradually becoming known ; the discovery
of the organisms responsible for the formation of
nitrates in the soil, and the conditions under which

these bacteria thrive, have thrown much light on the operations of tillage and manuring. Although it had long been known that certain plants seemed to enrich the soil on which they were grown, it was the discovery of the bacteria inhabiting the roots of leguminous plants, which are able to obtain nitrogen from the air, that has taught us how and by what these plants enrich our land.

All over the world scientists are at the present time engaged in investigating the problems connected with agriculture. The cross-breeding of plants and animals, the feeding of animals, the manuring of crops, the tillage of the soil, the improvement of pastures, the manufacture of dairy products, and very many other subjects are receiving the attention of hundreds of workers in Europe, America, and Australasia.

In the face of these facts, of which I have merely given the briefest possible outline, can the farmer ignore the benefits he has received from science? Many farmers, however, are apt to take their view merely from where they stand now, and forget that many of their common practices were unknown less than a hundred years ago. Because they always knew about superphosphate, cakes, analysis, and bacteria, and because they never knew the ravages of rinderpest and foot and mouth disease, they assume that these things ever were so.

There have always been a few pioneers in agricultural improvement, but it is rather remarkable that these men have seldom, if ever, been tenant farmers. We must remember, however, that till quite recently the farmer has been a man of practically no education, that the means of communication were bad; and it is possible that intelligent farmers may have brought about improvements of which the world never heard, till some one else wrote about it.

Although we must not forget that it was the Romans who attempted to invent some of our most modern labour-saving machines, and who cultivated nearly all of our most useful plants, including the turnip; and that it was the monks of the middle ages who kept agriculture alive in this country; yet, to Jethro Tull belongs the honour of being the great pioneer of the agriculture of recent times.

Born at the end of the seventeenth century, Jethro Tull received a University education, was called to the Bar, and after travelling a good deal, he settled down to attempt to improve his land, so that he might let it at a higher rent. He was perhaps the first to grow turnips as a field crop, and amongst the earliest to introduce clover into farm practice, but his name chiefly lives as the inventor of the horse-hoe and drill. In 1731 he published a work on " Horse-hoeing Husbandry," which, besides being widely read in this country, was translated into more than one foreign language.

Lord Townshend, better known as " Turnip Towns-hend," after retiring from political life in 1730, took up farming on his estate at Rainham, and his memory is kept alive, not by his political works, but by the fact that it was he who introduced the turnip and the four-course rotation into Norfolk.

Robert Blakewell, of Dishley, was the great pioneer of stock-breeding, and established flocks and herds famous all over Europe. From the hurdle-backed, long-legged, wool-bearing animals of those days he evolved the

Leicester sheep, and from the long-horned beasts of
burden he produced the first race of butcher's cattle,
the Longhorn.

The latter half of the eighteenth century witnessed
a great revival in agriculture. Enterprising farmers
came long distances to procure rams and bulls from
Dishley at extraordinarily high prices ; the growing of
turnips and clover demanded the enclosure of lands ;
this enclosure made the rotation of crops possible ; and
these improvements, carried out by farmers and landowners dotted sparingly about the country, made centres
round which the farming practice gradually improved.

When Arthur Young first began his famous travels
about 1766 he found much of the land covered with
wastes and commons, the arable land still unenclosed,
and largely cultivated on the "two crops and a fallow"
system ; but before his death the greater part of the
lands were enclosed and the rotation of crops had
become common. Arthur Young himself had, perhaps,
more influence upon the rapid movement in agriculture
than any other man the world has known. The son of
a country squire, farming because he preferred it to any
other occupation, it was not his practice that gave him
this influence, but his extensive travels and his remarkable and prolific writings.

Young's accounts of his travels were eagerly bought
and read by the agriculturist of the day, and they, together with the "Annals of Agriculture," which he edited,
had a far-reaching effect in bringing before others the
best practices of the day. He was made secretary to
the Board of Agriculture, and, although his wide experience of farming no doubt made him useful to that
body, he was always a farmer at heart, and was never

so happy as when viewing, or writing about, some improvement in farming.

By the end of the eighteenth century many men were devoting themselves to the improvement of agricultural practice, both in tillage and cropping, while the improvement in live-stock was particularly rapid. This improvement has been continued up to the present time, rapidly in the case of our scientific knowledge, and more slowly, but still more evidently, in the case of our practice. A hundred years ago science was endeavouring to find out the why and wherefore of practice; to-day science clearly shows us the lines which our practice ought to follow. We cannot afford nowadays to neglect any hint which may help us to cheapen our production or increase our produce. There is no doubt that the sciences are helping us towards this; engineering, physics, chemistry, physiology, botany, zoology, geology, and many other branches of science, are all adding something to our knowledge of the lines along which Nature works.

The knowledge of all these things will not necessarily make a successful farmer, but as no one is brought into such close contact with Nature as the farmer, surely to understand something of her ways is important. I have heard farming spoken of as a struggle with Nature, but the farmer who struggles with her must undoubtedly be beaten in the end. He must ever walk with her hand in hand, and he will then find that far from being the fickle goddess she is represented, Nature moves along certain paths, which he had failed to observe.

The processes by which Nature maintains the fertility of the soil may be assisted by intelligent tillage and cropping. The conditions under which certain plants

B

luxuriate may be copied in our practice, and we may invite the aid of Nature for the health and well-being of our live-stock.

No book can make a farmer, or make farming pay, but it may trace some few of Nature's paths, and thus become useful as a guide to farming. If this volume of mine shall add anything to the pleasures of farming by throwing a new light upon some difficult path, or by suggesting a method which may be used to aid Nature for our benefit, then its object will be fully attained.

CHAPTER II

Fertility

EVERY one will agree with me, I think, in considering the subject of fertility to be of the utmost importance to the farmer, and to deserve the first and fullest consideration. In selecting a farm, the farmer is always anxious to obtain one in a high state of fertility, and is willing to pay a considerable rent for it.

The efforts of all good farmers are directed towards maintaining and improving the condition of their land, but to do this intelligently, it is at any rate helpful to know what fertility is, and what are the factors that bring it about. If we were to ask a hundred farmers to name the factors which conduce to productiveness in a soil, I am certain that ninety-nine of them would begin with *muck*, and that quite half of them would end there. If we questioned them, however, we should find that their practice was very much wider than their theory, but that they had failed to connect many of their operations with the idea of fertility.

Take this example. No applications of farmyard manure would ever make a bog produce a crop of oats, but if it were thoroughly drained and well cultivated, it would probably produce several abundant crops without the application of any manure whatever. Now, it is evident that muck had nothing to do with the production of this fertility; but if we continued year after year to remove the crops, there would come a time when manure would become a factor in the fertility of that land.

The dictionary tells us that fertility means productiveness, and in this sense Nature tries to make the whole world fertile. The pond and river have their masses of water weeds; the bog or fen grows reeds, rushes, and horse-tails; the clayey soil, docks, coltsfoot, and creeping bent grass; the sandy waste, bracken and furze; and even the boulder on the mountain side will have its crop of mosses and lichens. All these plants grow naturally under the conditions of air and water that are most suitable to them, and where they are free from undue competition in the matter of food and light. The situations these plants select are therefore fertile, as far as they are concerned.

The definition of fertility, from a farming point of view, has to be circumscribed, and may be stated as that power which the land possesses of producing crops under cultivation. Even in this restricted sense it is a wide and complex subject, for although the factors of fertility are always the same, the means of bringing them about are both varying and numerous.

The *fellah* sows his seed in the flood-waters of the Nile, and reaps an abundant crop. The prairie farmer merely ploughs, sows, and reaps, and for several years at any rate grows good crops. The "cockatoo," or Australian squatter, gets his name because he merely scratches the surface of the ground; and there are lands in this country so naturally fertile that to keep them clean is to ensure good crops.

Except on such favoured Edens, the production of the conditions necessary to fertility requires much effort and skill; the clays are too wet, the sands are too dry, and the difficulties of cultivation are many.

Although the presence in the soil of an abundance of plant food is essential to the growth of a plant, yet it is seldom indeed that infertility is caused by the absence of the proper food substances. In every soil worthy of the name there exists enough plant food to grow dozens or even hundreds of crops, but the land may still be infertile, owing to the failure of the other necessary conditions, conditions which make the food available to the plant.

When a seed of any kind is placed in the soil, three things are necessary to bring about its germination. These three things are moisture, warmth, and air, and if any one of these is withheld the seed will fail to grow. The seed contains within itself a very small but complete plant, and a sufficient store of nutriment to enable this little plant to grow and push down roots into the soil and a shoot up into the air. The supply of moisture and warmth being sufficient, the leaves of the shoot will then be able to absorb carbonic acid gas from the air, and from this, with the assistance of the sunlight, to obtain the carbon it requires to build up its tissues. Besides supplying the water, which, together with the carbon

taken from the air, makes up about 97 per cent. of the
weight of a growing plant, the roots have to supply
a large number of substances taken from the soil in
comparatively small quantities, but without which the
plant would fail to live and grow in size. These substances are nitrogen, phosphorus, potassium, lime, magnesium, sulphur, chlorine, iron, sodium, and silicon.
By enumerating these substances in this way I wish to
emphasise an important point. We do not know exactly in what form these substances are taken up by the
plant, but we do know that they are absorbed by the
roots when combined together to form such substances
as nitrate of lime, sulphate of magnesia, or phosphate of
potash, and it makes no difference whatever to the plant
whether these substances are formed from rotten straw
and dung, or are supplied direct from bottles taken from
the shelves of a chemical laboratory. It is, however,
absolutely essential that a sufficient supply of these
elements should exist in the soil in such a form that
they are soluble in water, or in the weakly acid juice of
the roots themselves. Food substances which cannot
be dissolved in either of these ways are useless to the
plant for the time being.

The part of the root which does most of the work of
taking up food and water from the soil is an inch or so
just behind the extreme tip of the root. This portion is
densely covered with minute, white, silky-looking hairs,
which are in reality elongated bag-shaped cells, and into
these the water containing the soil food in solution is
absorbed, and is passed from them to other parts of the
plant. These hairs insinuate themselves between the
minute particles of the soil, and by exuding a slightly

acid juice they are able to dissolve from these particles
substances that would not have been dissolved by water
alone. As the root grows forward and branches in
every direction, new hairs are formed and the old
ones die off behind. Thus the soil is very thoroughly
searched for food when the conditions are favourable
to proper root growth, but it cannot take place where
the soil is hard and lumpy or sodden with water.

Air has been mentioned as one of the essentials for
the germination of a seed, for oxygen is just as essential
to the well-being of the living parts of a plant as it is
to animals. Most plants, however, possess no special
apparatus or part through which this oxygen is received,
but it is absorbed straight through the skin, so to speak,
wherever it is required. Now, as the roots of our farm
plants are alive, they must have oxygen, and therefore
oxygen must be present in the soil for them to absorb.
In badly tilled, sodden, or water-logged soils, the air is
excluded, and roots under such conditions become unhealthy, the root hairs die off, and the power of the root
as an absorbing organ is curtailed or ceases altogether.
Who has not observed the stunted, yellow appearance of
wheat in the water-logged furrows after heavy rains,
which is largely due to this cause? Certain plants,
particularly those which grow naturally in water or in
wet places, take in air by their leaves, and are provided
with special tubes down which the oxygen travels to
their roots. It is for this reason that such a plant
as horse-tail (*Equisetum*) can thrive in soils where farm
crops starve, while many other plants, such as creeping
bent grass (*Agrostis*), the creeping butter-cup and cinque-
foil (*potentilla*), manage to thrive in water-logged soils

by creeping over the surface and sending out large numbers of short rootlets which find their food in the top inch or two of soil. It is, indeed, a well-established fact that all plants grow naturally in those positions in which they have to put up with the least competition from other plants for food and light, and most plants have some special characteristic which enables them to thrive where other plants would fail. Our ordinary farm plants, were they growing in a state of nature, would be found to survive in those special situations in which they were able to thrive to the exclusion of their rivals. That this is to some extent the case is recognised by agriculturists when they speak of wheat and bean land, or of a turnip and barley farm; but were our farming operations confined only to those soils and situations most favourable to particular crops, a very small portion of our country would be tilled.

It is the business of a farmer so to cultivate his farm as to make the conditions those most suitable for the crops he is trying to grow, and the skill of the farmer in this respect is tested in two ways: first, by his being able to grow abundant and well-developed crops of various kinds, and, secondly, by his being able to do this in such a way that the crop grown will more than repay him for his outlay in producing it.

Fortunately the general conditions required by all our common farm crops are very similar, although there are differences in detail, but the farmer may aim at such conditions as will make his land more fertile for all his crops. It is of course possible to select for growth those crops which require the least modification of the prevailing conditions, and to do this properly is an evidence of skill and experience; but there are many cases in which success in the culture of a crop is determined by some slight difference in the method of cultivation, or by the addition of some manurial constituent which the plant finds a difficulty in obtaining from that particular soil.

That the mere presence in the soil of large quantities of the substances necessary for the growth of plants is not by any means the most important factor in fertility is shown by the following example. The soil of a field under experiment on my own farm was analysed, and it was found that an acre of the soil, 9 inches deep, contained 6300 lb. of nitrogen, 6600 lb. of phosphoric acid, 19,200 lb. of potash, and 27,000 lb. of lime. These are the four most important plant foods, and some idea of their relative quantity may be formed from the fact that a ton of average farmyard manure contains about 10 lb. of nitrogen, 5 lb. of phosphoric acid, 12 lb. of potash, and 15 lb. of lime. Now this land, although it contained these enormous quantities of possible food substances, only produced the miserable crop of 13¾ bushels of wheat per acre, nor did considerable dressings of dung and artificial manures increase the yield more than two or three bushels. It is evident that something very radical was wanted, besides mere food supply, to bring about fertility. As a matter of fact the soil was clay and the field undrained, so that it lay wet and cold all winter although the wheat was not killed, and in the summer the land became dry, hard, and cracked in all directions.

This is of course a very extreme case, but it shows that the crop depends more on the condition of the soil, and on the consequent condition or "get-atableness" of the food, than on its gross quantity. The conditions principally concerned in making the soil food available to plants are fine subdivision of the soil and a proper supply of moisture and air.

With regard to moisture, it seems that the proper amount of water in a soil should be from 15 to 30 per cent. of its weight, *i.e.* it should be slightly moist without being wet. Under these conditions it has been found that every minute particle of the soil is completely surrounded by a film of water which is constantly in motion and as constantly dissolving away some parts of the particle which may become food to a plant. As these soil particles, although often excessively small, are of irregular shapes, they have irregular spaces between them which are not filled up by the films of water but are occupied by air, and, as the films of water touch one another at the points of contact of the particles, so the air spaces communicate with one another in a well-tilled soil to a considerable depth below the surface.

When a root is growing in such a soil and the root hairs push themselves in between the particles, they are able to obtain all the oxygen they require, besides an abundant supply of water containing all the various substances dissolved from the particles. As the root absorbs the water surrounding the particles with which it is in contact, the films of water of course tend to become thinner in that place, but the same law of nature which holds the films on the particles will not allow them to be thinner in one place than another, and so water immediately flows to the thin place from other parts; and as the film of one particle is in contact with the film of another, water will flow in from the second particle to the first, and from a third to the second, and from a fourth to the third, and so on, until the whole of the water for a considerable distance is kept in a state of constant motion. Evaporation of water into the air is also constantly going on from those particles on the surface of the soil, and the water is as constantly flowing up from below to endeavour to supply that lost on the surface; so that when we have, besides the evaporation from the surface, many thousands of root-lets busy absorbing their supplies, the whole of the water in the soil to a depth often of many feet is kept in a constant state of motion.

Let us look for a moment at the beautiful working of this arrangement. As the water passes from one particle to another it is able to dissolve something from each of many different particles, phosphate of lime from one, sulphate of potash from another, and so on, until by the time it reaches the root it contains a collection of many of the substances required by the plant. Then, again, when the water travels upwards to the surface all kinds of substances are carried up which have been obtained from particles far down in the soil, and as the water evaporates these substances are left behind, and accumulate in the upper layers of soil, where they are within easy reach of the roots; or they may be left dry on the very surface. I have often observed patches of soil in dry weather covered with a thin layer of white powder, which is in fact a deposit consisting principally of sulphate of lime brought up from below and left on the surface by this means. When rain falls after this drying process, these surface accumulations are re-dissolved and washed down into the soil again, and the films of water on the particles, which had been gradually getting thinner, are now thickened by the new supply, and this thickening gradually travels downwards as long as the supply holds out. As these films thicken and occupy more space, a considerable quantity of the air in the soil is expelled, and when the water is again used and the films get thinner, fresh air flows in to take its place. In this way every shower acts as a deep breath to a well-tilled, well-drained soil, and even when an excessive quantity of rain has filled up the air spaces and saturated the soil, the surplus is soon drained away, leaving only films on the particles, and fresh air fills up the spaces.

When, however, the tillage is bad and the drainage imperfect, this surplus water remains, making the soil, for long periods at a stretch, sodden and water-logged. As the water cannot get away through the soil it has to evaporate from the surface, and in the bright days and keen winds of spring the soil becomes excessively cold owing to this evaporation. I have often observed a thin layer of ice encrusting wet places in a field long after the rest of the soil began to get warm and dry, because the heat of the sun's rays is used up in evaporating the water instead of warming the soil. It is for this reason that spring growth is so much later in starting on a damp clay than on a dry and sandy soil.

When a soil is water-logged, that is, when the spaces between the particles are filled with water, air is of course excluded, and there is none of that circulation of water so essential to the solution of soil foods. The absence of air means that the plants soon become unhealthy, the root hairs die off, and eventually the roots of themselves will rot. When the decomposition of vegetable matter takes place in the absence of a proper supply of air, certain acids are formed which make the soil sour and unfavourable to healthy plant life. So well is this known to practical men that to say a soil is sour is synonymous with saying it is wet.

When such a soil as this does become dry, it sets in huge blocks, between which gape enormous cracks, many I have observed being from two to three inches wide and as many feet in depth. When ploughed it becomes an expanse of adamantine lumps, over which the harrows jingle and the roller skips with a merry rattle, but with the minimum of effect. Under such circumstances the roots of plants fail to obtain their food, being unable to penetrate the lumps, and such a field will suffer very quickly from drought. During several of the late dry summers I have frequently observed the lines of the drains marked out in vivid green while the rest of the crop has been dwindling away for want of rain. Efficient drainage and deep steam cultivation are the principal means of bringing such land into fruitful condition, and I know too well from bitter experience the sadness of attempting to cultivate it where this has not been carried out. Dressings of lime will also quickly‘ correct the sourness of such land, and will bring about that neutral or slightly alkaline condition so beloved of most plants.

In very light and sandy soils, the chief difficulty in their cultivation lies in getting them to retain enough moisture to last the plants from one shower to another. The particles are comparatively so large that the rain runs straight through the spaces between them and quickly reaches depths at which it is useless to the plants, while the quantity of water retained within reach of the roots is less than is the case in a soil composed of a larger number of smaller particles. The spaces between the particles are larger, and therefore contain more air; the circulation of the water is more rapid and the evaporation greater in proportion to the amount of water present; while the soil food which becomes dissolved is very liable to be washed down out of reach of the plants, and consequently such land tends to become very poor. These effects can be minimised to some extent by the compression of the subsoil, and by the addition of large quantities of vegetable matter. These very light and sandy lands are principally used for market gardening, for it is only crops of that nature that will pay for the enormous dressings of dung generally used to keep up their fertility, both by retaining moisture and supplying food.

Most soils contain extremely large stores of reserve plant food, of which only a small proportion becomes available for use at any one time. In land in a state of nature or covered with permanent pasture, and occasionally when very highly farmed, this reserve tends to accumulate in the top soil year by year. The particles of soil, being largely minute fragments of rock, contain in themselves enormous stores of the mineral matters required by plants, and as these are slowly dissolved they are built up into the substance of the plants. When these plants die the whole of this substance is left as vegetable matter, either on the surface or as roots in the soil. This vegetable matter, besides containing phosphorus, potassium, lime, &c., possesses nitrogen, which was originally obtained from the air, and about which I shall have more to say later on. Thus the top soil tends to become richer in plant food, obtained from the subsoil and from the air, and as decay takes place these various substances are again made ready for the use of a new generation of plants.

The chief factors in determining how much of this reserve food shall become available for plant use are the relative supplies of moisture and air, so that in a clayey soil a comparatively small proportion will become available in any one year, and a larger proportion added to the reserve; while in a sandy soil the greater quantity will become available, either to be used by plants, or to be washed away by rain, and only a small proportion will be added to the reserve.

In ordinary arable lands, where the crops are removed and only the manure made on the farm is returned to the soil, this reserve tends to get less, and on well-drained and thoroughly cultivated farms a large proportion of the reserve will be made available, and increased crops will result. No intelligent farmer will, however, expect to eat his cake and have it; and so, if he forces Nature to unlock her stores for his benefit, he must see that he prevents those stores from becoming unduly low, and maintains a system of judicious cropping and manuring. It is, I am convinced, only by making an intelligent use of these stores that it is possible to make ordinary farming pay, and it is possible to make such use of them without in any way robbing our soil; indeed, by making these stores soluble and available, we are contributing towards that true fertility which we are seeking.

By adding manures to undrained, unsuitably-tilled soils, we are only adding the greater part of this manure to the already enormous store of unavailable reserve; but if we have taken the means to bring the reserve into use, then the manures we add will bring about a due increase in fertility. It is necessary to remember, however, that every soil does not yield up its stores in an available condition, exactly in the correct proportions demanded by the plants.

Some one or more of the necessary soil foods may be in excess of the requirements of any possible crop; another constituent may fall far below that required for even a moderate yield. In this case it is our duty to find out by experiment what that constituent is, and to add it to our manure; for it is an unalterable rule that, just as the strength of a chain is that of its weakest link, so the size of the crop grown is determined by the amount of the least available constituent present. Fortunately for us, all soils are able to provide, in an available condition, abundance of magnesium, sulphur, chlorine, iron, sodium, and silicon: but there are a few soils that fail to supply a sufficient quantity of potash for the requirements of all crops; there are rather more that fall short of the supply of lime necessary to high fertility; there are a large number that fail to provide an abundance of phosphates; and nearly every soil requires the addition of readily available nitrogen at frequent intervals. I shall enter fully into the best means of applying these substances later on, but for the present I wish to confine myself to the influence of their availability on fertility, and I have endeavoured to show that thorough tillage and the presence of air assist greatly in dissolving the constituents of the mineral matter of the soil.

The methods by which the animal and vegetable remains in the soil become again available for plant use now demand consideration, and this, together with the means by which plants obtain their supplies of nitrogen, forms one of the most fascinating and interesting studies in the economy of the soil, but of which the demands of space will compel an all too limited description.

The upper layers of every soil, of whatever kind, are swarming with myriads of minute organisms, of which those known as bacteria form the greater part. A single cubic inch of soil will contain them in millions, and the better tilled, the richer and more perfect the condition of our soil, the greater become the number and the activity of the bacteria. These bacteria are of many kinds, each of which is intent upon obtaining a livelihood by its own method, some of them being dependent upon the work of the others for the means to subsist.

Whenever the organic remains of any animal or vegetable are added to the soil, these are immediately invaded by myriads of bacteria, that proceed to feed upon them, and in doing so break down the complex organic substances of which they are composed into large numbers of simpler compounds. During this process of breaking down, which we describe as putrefaction, different gases are produced, some of which we can detect by their odour; substances are formed containing the phosphorus, potassium, and other components of the body; but most important and valuable of all is the formation of compounds of nitrogen in the form of ammonia.

As soon as these ammonia compounds come in contact with the soil, they are seized upon by another bacterium which is able to make use of them, and as

C

the result of its work, the nitrogen of the ammonia is combined with oxygen and left behind as a substance called nitrous acid.

Now if this acid were left in the soil as it is, not only would the land become sour, but it would cause such conditions that the bacteria would be unable to live in it. This danger is prevented, however, by the presence of chalk (carbonate of lime) in the soil, for the acid combines with the lime, forming a substance, nitrite of lime, which is no longer acid.

This compound is again of use to another species of bacterium, which for some good reason of its own tacks on to it another atom of oxygen, and turns it into a compound of nitric acid called nitrate of lime.

The process by which ammonia is turned into nitric acid is known as *nitrification*, and it is of the utmost importance to the plants and to us, for we know that green plants can take up nitrogen from the soil in no other form than that of a nitrate of some kind, and usually as nitrate of lime.

This being the case then, it behoves us to do everything in our power to encourage the growth and activity of these useful bacteria. Fortunately the most suitable conditions for them are also those beloved by our farm plants, namely, moisture, warmth, and air, while of course they require the necessary decayed organic matter to work and feed upon, and an adequate supply of lime in the soil.

Strangely enough, there exists in the soil another class of bacteria, whose function it seems to be to destroy the work of the useful kind we are so anxious to encourage, although in reality they are only doing their part in the working of the scheme of Nature. In heavy and water-logged soils where air is excluded the organic remains of vegetation tend to accumulate year after year, and except on the very surface the ordinary decay cannot take place for want of air. It is in such a situation that this class of bacteria are able to thrive, for when air is not available, they are able to obtain the necessary oxygen by breaking down certain compounds containing it. The most readily available compound for them seems to be the nitrate of lime formed by our friends the useful bacteria; and when they have extracted the oxygen they require the compound is no longer a nitrate, and the nitrogen it contained goes off into the air as a gas, only to be added to the tons already existing there in a useless state. This destruction of nitrates goes on in heavy wet soils, containing undecayed vegetable matter to a much greater extent than is generally supposed; indeed, I have found by carefully conducted experiments that under ordinary field conditions these bacteria were able, in the presence of thirty tons per acre of strawy manure, completely to destroy one and a half hundred-weights of nitrate of soda, and under the most favourable conditions in the laboratory I have seen nitrate of soda equal to a ton and a half per acre destroyed in four-teen days. It is lucky for us that the very conditions so suitable to green plants and to the useful bacteria are those under which these destructive bacteria cease to thrive, and there is no doubt that in well-tilled and thoroughly aerated soils the presence of these *denitrifying* bacteria can be ignored.

There are very many other kinds of these invisible

organisms which are of importance to agriculture, and I must mention one more class of bacteria while we are on this subject.   These are the nitrogen-collecting class which live in the roots of leguminous plants, *i.e.* of such pod-bearing plants as beans, peas, tares, clovers, &c.   These bacteria exist in the soil, and when a plant of this kind begins to grow they enter into its roots, forming on them little lumps or nodules, in which they live and multiply.   In some way or other these bacteria are able to obtain nitrogen from the air, which, after they have used it themselves, no doubt, becomes available for the use of the host plant.   Although these bacteria live on the juices of the root, the plant is really benefited by their presence ; and when this host plant dies, it leaves, *as a gain to the soil,* the nitrogen it obtained through the bacteria, from the air.

How marvellous is this unceasing round of life and how few the people who regard the soil as other than inanimate dirt !   Surely the tilling of a garden or a field must become a double pleasure when we think of the teeming population of the soil, all busily engaged in earning their own living and by their efforts helping us to earn ours.

I hope that I have now been able to show that fertility in a soil depends on many other things besides dressings of manure, however useful they may be.   We have been able to see that the inherent capabilities, or natural tendencies of a soil to fertility, depend very largely on its permeability by air and moisture, the quantity and solubility of its mineral plant food, and on the quantity of its organic matter.   We shall, I think, be able to get some useful information as to their characters by briefly considering how soils are formed, and of what they are made.

The geologist would probably tell us that the soil is a layer of finely disintegrated rock particles containing a certain amount of organic matter.

To the farmer, however, this definition is rather too wide, as it may take in any depth, however great ; so that I think we shall not be far wrong if we define the soil agriculturally as that portion of the surface layer which is usually turned over by the plough or disturbed by the ordinary acts of tillage.

The subsoil will then be that portion of the upper layer which underlies the tilled portion and into which roots penetrate. Geologically, however, the subsoil is that portion of the rock underlying the soil, and which is in the process of breaking up into soil. The two definitions are necessarily somewhat different because the soil and subsoil of the farmer are more restricted than those of the geologist.

We can get a great deal of information as to the composition and needs of a soil by studying how, and of what, it is made. Every kind of substance of which the earth's crust is made is known to the geologist as a rock, and, no matter how hard and resistant these rocks, if they are exposed to those influences we call the weather, they are all in the process of being broken up and eaten away. This process can be seen going on in the stones of any old building, and the same thing is going on in fields, hills, and cliffs all the world over.

The influences included under the name of "weathering" are principally those of the air, rain, and change of temperature.

The air supplies oxygen and carbonic acid gas, which act chemically upon the rocks; while the rain not only breaks up soft rocks by its beating action, but carries into the crevices of harder rocks a great deal of oxygen and carbonic acid. By means of these gases the rocks are oxidised and dissolved, often very rapidly, as in the case of limestone rocks. Change of temperature, again, brings about much fracturing by unequal expansion and contraction, while the freezing of water in cracks is accountable for a great deal of disintegration.

The result of all this is ultimately the formation of very fine particles of rock. Amongst these particles certain lowly organisms, such as bacteria, lichens, and algæ, begin to grow, many of which doubtless obtain their nitrogen from the air, while they can obtain their mineral matter from the particles of rock.

As these organisms die and leave their nitrogenous remains, higher plants, such as mosses, make their appearance, and finally, feeding upon the remains of these, little rock plants and grasses are able to grow. In this way, little by little, the growth and decay of these various plants adds organic matter to the particles of rock, and a soil is formed.

Very many soils are thus formed by these agencies, and are said to be formed *in situ*, for of course the subsoil and rock to which they belong will be found underneath them. Naturally clay, lime or sandstone will give rise to clayey, chalky or sandy soils, but a clay may also be formed from several different kinds of rock.

A large number of soils, however, are made from particles which have been moved from the place where they were originally formed. Such soils are called transported soils, and they may be either lying on an entirely different subsoil, or the subsoil may have been transported as well. In these cases the underlying rock gives no clue to the composition of the soil.

The transportation of soils takes place principally by the carrying action of running water, though some have been carried by ice, and a few by the wind.

Wherever rain falls on a sloping surface, particles are washed downwards to a greater or less extent according to the amount of rain and the slope of the surface, so that there is gradually formed in the valleys a soil consisting of a mixture of particles from various parts of the hills. The result of this is to produce a rich soil at the expense of the upper parts of the hill.

Again, a very large quantity of particles are carried into streams and rivers, and are thus transported a long way from their origin, often out to sea. In times of flood, however, these particles are deposited on the flooded land, and there is thus formed on the river flats a considerable depth of alluvium, which is usually the richest of all soils, as it contains a mixture of particles from many sources, and a considerable quantity of vegetable matter.

If we examine a soil roughly, by shaking a little up in water, we can easily make out differences in its materials. There may be some stones or gravel, then a considerable quantity of coarse sand which will quickly settle, and a smaller quantity of very fine sand which settles down more slowly, leaving suspended in the water for some time the very fine materials which we may call clay and silt. There will also be interspersed a quantity of vegetable matter.

The weight of an acre of soil 9 inches deep and quite
dry is about 3,000,000 lb. or 1339 tons, and when
moist it would weight considerably more.

An actual analysis of clay soil when dry gave the
following figures, omitting fractions :—

Stones and Gravel	.	4 per cent.	or about	54	tons.
Sand	.	53 ,,	,,	710	,,
Very fine Sand .	.	12 ,,	,,	160	,,
Clay and Silt .	.	22 ,,	,,	295	,,
Vegetable Matter	.	9 ,,	,,	120	,,

When moist this soil would contain in addition from
two to three hundred tons of water per acre.

The amount of true clay, known chemically as hydrated
silicate of alumina, is really very small even in the most
clayey soils, probably not more than 5 or 6 per cent.,
but owing to its exceedingly sticky nature its effects on
the consistency of a soil are very great. A friend of
mine was once walking over a farm, and looking rue-
fully at his boots, he remarked to one of the men that
"they had some uncommonly sticky mud in these
parts." "Mud, you call it," said the man ; "we don't ;
we calls it 'idrated silicate o' 'lumina." They had had
some extension lectures in the village during the winter.

It is to this true clay that heavy soils owe their
adhesive character, but it is the silt of impalpably fine
particles that makes them so impervious to water, and it
is these fine particles that contain most of the mineral
food of the soil.

When treated with strong acid, the soil just mentioned
was found to contain the following quantities per acre of
the substances required by plants :—

Phosphoric Acid	9,600 lb.
Potash	21,900 ,,
Chalk	46,500 ,,
Lime in other forms	24,900 ,,
Magnesia	15,000 ,,
Soda	8,000 ,,
Sulphuric Acid	9,900 ,,
Iron Oxide	190,800 ,,

The vegetable matter of this 9 inches of soil also contained 6000 lb. of nitrogen per acre, while it was found that except in nitrogen the next 12 inches of subsoil were very nearly as rich in possible plant foods as the upper soil.

The comparatively large sandy particles contain very little plant food, being principally composed of quartz, but their usefulness in keeping the soil open makes them of great value. Few clay soils contain less than 25 per cent. of sand, and a soil may still be clayey even if containing 60 per cent. or over.

The sand of a soil then is principally advantageous by reason of its mechanical effects. The particles of sand will never cohere or stick together, therefore a sandy soil works easily and is called a "light" soil, although it is really heavier than clay. The largeness of the particles makes sand always porous, and it will retain very little of the water that falls upon it. This largeness of the particles prevents water from rising from below by capillarity to any great height, and it becomes dry very quickly.

Clay—that is, the impalpable powder which we call clay —is in many ways the opposite of sand. Its particles are extremely fine, and when moist cohere or stick together very strongly. It is scarcely permeable to water; indeed, when worked about so as to expel the little air it contains, it becomes "puddled," and is practically quite impervious. It is very retentive of water, and when dry is able to absorb moisture from the air. Water rises in it very slowly, but will rise to a great height.

The vegetable matter of the soil, when it is partially decayed and assumes a brownish colour, is spoken of as humus, and as such it is a remarkably important constituent of the soil. It contains all the constituents of plants, and is the principal source of nitrogenous food.

Humus has a spongy texture, and has enough cohesion to hold a sandy soil together, but not enough to make it sticky and heavy, indeed it makes clay easier to work. It is very porous, and it is able to absorb and retain a great deal of water. Its sponginess enables it to raise water to a great height with great rapidity, so that it helps both sand and clay in this respect. When dry it is able to absorb moisture from the air to even a greater extent than clay.

Chalk is a large constituent of many soils, and should of course be present to some extent in all. Its influences on the condition of the soil place it about midway between sand and clay in that respect.

Stones are of little value to the soil as sources of plant food, but they are often valuable when not too numerous in keeping the soil open, and in scouring the tillage implements, particularly in preventing the fine earth from sticking to the breasts of ploughs.

It will be apparent from the differences in the properties of the various soil constituents that the soils will vary much in their properties according to the proportions of these constituents present, and the cultivation and manuring of such soils will vary too.

Our knowledge of the science of the soil is very far from complete, but one cannot help thinking that if some of the old pioneers of agriculture had known even as much about it as we do, they would have made far more use of it. We have seen that the stores of mineral food in the soil and subsoil are sufficient to supply our crops for hundreds of years, if we can only manage to make them available as required. We can capture the nitrogen of the air, and add it to the soil by the cultivation of the proper crops. We can add the ingredients which may be deficient in the top soil by dung or by chemical manure. All this we can do to supply plant foods in abundance; but if the condition of the soil as to aeration and moisture is not such as is suitable for the healthy action of plant roots, and for the active multiplication of soil bacteria, our efforts towards fertility will be largely in vain.

That fine old farmer, Jethro Tull, in an age when men had hardly ceased looking for the philosopher's stone, came very near to discovering the true sources of fertility. His experiments with plants were most extraordinary in their advance on anything of the kind ever done before, and the conclusion he arrived at was that thorough and constant tillage was the means to fertility. To this he devoted himself, and by planting his wheat in wide rows, and by thoroughly cultivating the spaces between with his newly-invented horse-hoe or grubber, he was able for years to grow successful crops. Where he failed, however, was in supposing that, with thorough tillage, the soil required no occasional assistance in the form of manure.

The Rev. S. Smith, of Lois Weedon, grew very

abundant crops of wheat for many years on the same land by planting the wheat in strips about three feet wide, with an unplanted strip of similar width between, which was kept well cultivated throughout the summer, the wheat and fallow being alternated each year. Many other important experiments might be quoted to show how large a place deep and thorough cultivation holds in soil fertility. One cannot help being struck, on reading Bailey, King, Roberts, and other American agricultural writers, by the great importance they attach to thorough cultivation; deep, to provide abundant space for healthy root action, and shallow, between the plants during summer, to prevent the escape of moisture by evaporation.

Lawes and Gilbert, in their classic experiments at Rothamsted, have been able to show, amongst much other invaluable information, that good crops of the same kind can be grown for nearly sixty years on the same land by the use of chemical manures alone. Except in the case of clover, there is nothing in the appearance or yield of the crops to show that this cannot be done with success if required; indeed, Messrs. Prout, of Sawbridgeworth, father and son, have cultivated their farm on this system for over forty years; and although everything goes off and no manures are used but chemical manures, I have seen splendid crops there.

The Prouts have grown their crops to some extent in a rotation, and there is no doubt that their farm has paid; but where many of the pioneers have failed in practice has been in allowing their enthusiasm for an idea to run away with their judgment.

Mechi undoubtedly demonstrated the wonderful effects

of cultivation and drainage on the Essex clays by the production of splendid crops; but when his enthusiasm for drainage caused him to begin replacing his $2\frac{1}{2}$ feet drains by those 4 and 5 feet deep, the expenses got the better of him.

Tull was accused of drilling away his fortune; Young, carried away by his thirst for information, is said to have lost his money by making three thousand fruitless experiments on his Suffolk farm; and doubtless many of the wealthy pioneers, whose practices are in common use to-day, lost money by their farming.

Nor is this loss to pioneers peculiar to agriculture, for have we not all heard of fortunes lost in the endeavour to invent some machine or apparatus, and of the individual who, striking the missing idea, reaps the fortune which his predecessor had lost? There is always this tendency to let enthusiasm run away with caution, but we ought to be able to profit by the mistakes and successes of others. Unfortunately the farmer usually fails to distinguish the dividing line between profitable improvement of practice and reckless expenditure, and so condemns a whole system which may in itself be excellent.

Farmers as a rule are decidedly conservative, and I do not blame them in the least; but what I have had reason to blame some for, is their direct and active hostility to any innovation. Their conservatism arises from an instinct of self-preservation, a determination to stick by that which they have found to produce fairly good results, and not to take the risk of methods of which they know nothing; but hostility arises in every case from ignorance. The intelligent old-fashioned farmer is one of the most delightful of fellows, and there is nothing I enjoy more than a prowl round his farm with him. He has been about the country a bit, and still occasionally visits the Smithfield or Royal Shows. He has heard of this or read of that, and has no doubt that such and such a method is a good one, but he is too old to change now. You cannot help agreeing with him, dear old boy; and how he takes you back to what he calls the "good old times," and tells you tales of his father and grandfather that are better than a book!

His methods of culture may not be based upon labour-saving principles, but look at his wheat and his barley, such wheat and such barley, well cultivated and clean as a garden; it may not be grown at a profit, but it is fine nevertheless. His sheep—now those *are* sheep; giants, with backs like the seat of a sofa: he never sends to market the half-fat, nine-stone babies of modern times. His cows and his calves are excellent, first class; and his pigs! such mountains of fat, just as we used to see them at Islington thirty years ago. How delightfully old-fashioned are the surroundings of his home; the windows of the ivy-covered house look straight into the well-littered yards; the little patch of lawn, with its circular beds of sweet-smelling flowers; the rows of patriarchal gooseberry and currant bushes; and beyond these we can catch glimpses of the luxuriant beds of cabbages, onions, and potatoes for the household use.

How can we resist his ever-ready hospitality, especially as through the open door of the rose-covered porch there comes that delightful and indescribable odour of new-baked bread? The parlour we will have none of; the house-place, please; and so we follow our host into the great kitchen sitting-room, spotless in its

cleanliness, and occupy one of the cosy high-backed arm-
chairs which stand facing the great open hearth. His
wife, good old soul, soon loads the table with every
kind of delightful home-made viand, and while we chat
away we can observe the chimney corners hung with
their huge flitches and hams and bundles of dried herbs;
the couple of guns over the mantelpiece, the fox's mask
on the wall, and the silver cup on the sideboard, each
of which gives rise to stories of olden times. We must
not stay too long, however, for this is a busy household :
its work begins at five o'clock in the morning; there is
a ceaseless round of healthful and pleasant duties for
mistress and master; but its duties end with a cheerful
hour by the fireside, and by nine they are ready for
rest.

I have wandered away from my subject, however, in
thus speaking of a type of farmer which is necessarily
becoming scarcer and scarcer; but in every district
there are to be found some of the older farmers whose
methods show a marked superiority over that of their
neighbours. These men may know nothing of agricultural science, but they are always men who have
observed, read, and thought, and the advice and opinion
of such men is of the greatest value to the novice, however
highly versed in science he may be. Let me tell the
young farmer, fresh from the Agricultural College, that
if he possess such a man for a neighbour he is fortunate
indeed, for his example and precept will keep him from
many an error.

The reproach of the farmer as a "stick in the mud"
is, I am convinced, fast being removed. The old-
fashioned farmer, uneducated, unread, whose know-

ledge of agriculture was confined to what he saw in his own parish, would of course never improve, but he is fast dying out. This ignorant class is now only being kept alive by the sons of small farmers, too ignorant to recognise the benefits of education, either for themselves or their children. It is this class which the agricultural schools, both of this country and of North America, find it so difficult to reach. On the Continent much is being done by co-operative societies and indirect education, and in Denmark so successful has this been that there is now a general demand for agricultural teaching among the peasantry.

The intelligent farmer is always diffident as to the superiority of his methods, but the implements, manures, and feeding-stuffs he uses show that he has endeavoured to keep up with the times. What he too often lacks and knows he lacks is the education, the knowledge which will enable him to use these things to the greatest advantage in his practice. The education every farmer requires is some knowledge of the sciences bearing upon farming, for whether he realises it or not, every operation of the farm is the practical application of certain scientific principles.

Up to within a very few years ago the use of a certain number of chemical fertilisers was supposed to constitute scientific farming, at least that was the idea conveyed to most people by the term "scientific agriculture." We now realise that the term "scientific agriculture" has a much wider application, for all farming, however common and everyday it may be, is but the endeavour to turn certain laws of Nature to our own use. If, then, the farmer can have some definite, some scientific knowledge of these laws, will he not be better able to adjust his practice to varying conditions than he whose knowledge is confined to those experiences with which time has made him familiar?

In this chapter on fertility I have tried as briefly as possible to set out some of the laws which underlie the growth of farm crops, and I contend that it is the practical application of such principles to the tillage of the soil that constitutes the beginning, at any rate, of scientific agriculture. Will the delight which nearly every son of Adam feels in the cultivation of the soil be any way lessened by the fact that his cultivation is guided by his knowledge of Nature's laws? The possession of a square mile of Brazilian forest, resplendent with all Nature's luxuriance, would never give us the delight, the subtle pleasure which we feel in contemplating a little bed of cabbages grown by our own skill and care in an erstwhile wilderness.

It is just the same with the farm. Were it possible to find a piece of land so fertile that to sprinkle it with seed was all that was required to secure a crop, would that land provide us with delights other than those of the money-grubber, "the man with the muck-rake"? No: the pleasure of farming largely consists in seeing Nature blossom forth in that particular place and manner as the result of our own knowledge and skill. To drain the water-logged field, to manure the barren soil, to plough, sow, harrow, and roll, and as a result to produce increased crops for the food of man and beast, is not the least among the attractions of farming, and occasional failure only seems to spur us on to increased efforts. The feeling that Nature is bound sooner or later to respond to his efforts seems to have had something to do with the bulldog pertinacity with which

D

the British farmer has stuck to his land through all the bad times. He has hoped against hope; but the man who has been able to bring about the greatest improvement in his farming in the most economical manner has usually been the most successful. This economy is, as a rule, the result of making intelligent use of the experience of others, experience of practice, experience of science.

Many of the agricultural pioneers started upon paths which must have led them to the greatest possible success, had they known something of the science underlying their practice. Nowadays we are beginning to understand what it is we are really trying to bring about when we drain, or plough, or manure, and we are therefore in a better position to reason out the most effectual and most economical method of doing it. Such a method must undoubtedly pay the best: will it be any the less interesting?

CHAPTER III

The Improvement of Soils

IN the consideration of this subject there seem to be three fairly distinct types of improvement to which our knowledge of the science of the soil may be applied. Firstly there are those operations which have for their object the amelioration of the soil during a fairly long period, such as drainage, clay-burning, liming, claying and green-manuring; then there are the more temporary operations, classed as tillages; and lastly manuring. I hope to be able to say a little about each, taken in this order.

Draining is one of the most important and one of the most expensive operations of the farm, and it is sad to reflect that although there are thousands of acres which will never be fertile till they are drained, yet several hundred thousand pounds have been thrown away in

the past in draining land that never required it, and
in draining in a useless manner land that required it
badly. The fierce controversy which raged during the
forties, fifties, and sixties of last century, as to the
respective merits of deep and shallow drainage, showed
that the subject was never properly understood, or the
opponents would have seen that each might be right or
wrong according to the circumstances of the case; and
these men, and many others since, have failed to grasp
the fact that the differences in soils are so great that
what may be excellent treatment for one may be quite
wrong for another.

The object of draining should be to remove stagnant
water and to prevent it accumulating in or upon the
soil. Remember, we only want to prevent water standing; we do not want to dry the soil, or unduly to
remove the circulating moisture which performs such
an important part in the soil economy.

When rain falls upon a porous soil the water should
pass downwards through it, leaving behind in the soil
only sufficient to coat all the soil particles with their
proper films of moisture. The water that has passed
through the soil will, if the subsoil is also porous,
continue to descend, moistening the particles as it goes
until it is all used up, or until a non-porous layer
somewhere in the depths below arrests its progress.
When it has reached this impervious layer the water
will accumulate on it, saturating the pervious stratum
above to a greater or less depth. The top of this
saturated layer is spoken of as the water-table, and if
rain continues to fall, the surface of this water-table
will rise nearer and nearer to the surface of the soil.
Suppose, now, we dig a deep hole or a well in such a

position : no water will flow into the hole while we are passing through the subsoil and stratum in which the particles are only coated with films of water; but as soon as we pass into the saturated part, water will commence to flow in and fill the hole up to the height at which saturation ceased—that is, up to the surface of the water-table.

Now, if after a winter's rain we dug such a hole and found that no water stood in it until we had reached a depth of, say, 6 or 8 feet, we might safely conclude that this water could result in no harm to our plant roots, and such a soil did not require draining. If, on the other hand, water stood in the hole at a depth of 2 or 3 feet from the surface, then there would be danger of harm to our plants, and draining would probably result in benefit. If it were possible for us to reach the impervious layer and make outlets of some kind, we could prevent the mischief from occurring entirely, but this of course would be impossible except in a very few isolated cases, such, for example, as a saucer-shaped depression in the clay filled with gravel. These pockets of gravel, as they are called, often form wet patches when they rest on a clay substratum, and a large patch can often be drained by a single deep drain. All we can attempt to do in the majority of cases is to prevent the water-table rising too near the surface, and this can be done by putting in drains, into which the standing water will flow as soon as it reaches a certain height. This is the ideal case, and here pipe drains laid in at 4 feet deep, with intervals of from 20 to 30 yards between, would probably effect our purpose thoroughly. In soils which, though porous, are less so than the ideal one, the drains must be closer together, for it must be remembered that between the drains the water has to flow sideways to reach the pipes, and as this is a slow process, the water-table tends to become heaped up between the drains, and may remain there sufficiently long to do harm.

Much information may be gained by digging a series of trial holes about 4 feet deep at intervals of, say, 11 yards. If water stand in these, then draining is probably required, and we may find out something about the distance we may have between the drains by letting off the water from one of the holes so that it can flow away. If, after this has been running for a few days, the level of the water has fallen in the holes 11 yards away, then our drains may safely be laid 22 yards apart, and if a further hole is effected then we may go 30 or 40 yards apart, as the case may be. Care must be taken to prevent water flowing off the surface into these trial holes. We must remember also that a drain 3 feet deep will not drain so wide an interval as one 4 feet deep; and the depth and number of our drains have to be regulated by the relative porousness of the soil, for that which we have to do is not to carry away the water as quickly as it falls, but to prevent it accumulating in the subsoil in such a manner as to be injurious to our crops.

This, then, is what we may call the theory of draining in porous soils; but we shall find, I think, that the case is very different in clay soils.

In the first place, clays are not naturally porous, and it is they which usually form the impervious layer, causing the water to stand and form a water-table in a porous stratum. This being so, the water-table must necessarily be at the surface of a clay soil; and so it practically is, for although the soil which is tilled is porous, and the subsoil, which may be disturbed by deep cultivation or is pierced by roots and worms to a little depth, may be slightly porous, yet, as soon as it reaches the undisturbed clay, water penetrates no farther by gravitation. For this kind of land, then, all our theories which apply to porous soils are absolutely useless. In the place where I am now sitting we could dig a trial hole 100 feet deep without finding enough water to wet our boots, but twelve hours' rain would cover the undrained fields with pools. An old and very commonly used plan of getting rid of the water is to provide open furrows at intervals, down which the water can run, and be conveyed to the ditch. This does some good by taking away the water which would otherwise lie on the surface; but where the cultivation is imperfect, the subsoil can retain very little useful moisture, and such land is either soddened with water after rain, or suffering from drought after a week or two of sunshine.

It is without doubt essential that deep and thorough cultivation must go hand in hand with drainage on clayey soils, for although we remove the surplus water, we must have the subsoil capable of retaining sufficient moisture to carry the crops from one rain to another. If, then, we cultivate deeply and increase the moisture-holding power of our subsoil, it stands to reason that surface furrows will not remove the standing water from this deepened subsoil. In many parts of the country the old-time farmers used to plough their heavy soils in high-backed lands or ridges. Every one who has travelled through the Midlands must have seen them, and any of my readers who have hunted in the grass country will no doubt be acquainted with the delights of riding over them. Some of these may be able to sympathise with my reluctance to occupy a chair, when as a lad out for a ride with the Pytchley I lost my stirrups and was run away with over what seemed to me miles of these awful ridges. Well, the farmers who threw up those ridges were on the right track, for not only was the bulk of the ridge made of moved soil, and therefore porous and capable of retaining moisture, but the furrow was well below the pervious part, and capable of draining it thoroughly. I am indeed a great believer in high-backed lands of a moderate height, and have laid out many fields of them myself, but I was very much struck to see them being strongly recommended to the farmers of America by Mr. I. P. Roberts, Professor of Agriculture at Cornell University, in his book on "The Fertility of the Land."

The old-fashioned high-backed lands were seldom ever straight; indeed, they more often resemble an elongated letter S. The reason for this peculiar conformation seems to have puzzled many a wiseacre and called forth a number of ingenious and amusing explanations. Even Rider Haggard in his "Rural England" quotes the well-worn fables, but the reason seems to me to be obvious. Did you ever see, after a heavy rain, a little heap of sand and soil at the end of a furrow, where it had been carried down and left by the running water? Now these old boys knew better than to have their soil washed away like that, and one finds that the greater the slope of the field the more like an S the lands became, so that the water should run away too slowly to carry the soil with it.

I can state from experience that lands 11 yards wide and 2 feet high in the middle, or lands 15 yards wide and

3 feet high, present no obstacles to binders or other modern machines; but our fathers, when reaping-machines and drain-pipes became common, levelled the old-fashioned lands, straightened the furrows, and perhaps laid in drains at regular intervals. The result has been that in many cases these fields are now so wet and sodden that they fail to be worth cultivating. The fact of the matter is that, in a real clay, pipes fail to drain the land to any appreciable extent, for, owing to the impervious nature of the subsoil, there is no lateral movement of the water through the subsoil into the drain, and therefore the only water which is taken away is that which descends directly through the ground moved in putting in the pipes. Even this ceases after a time where the drains are laid too deeply. In such soils drains should never exceed 2 feet 6 inches as the average depth; and at least 12 inches of broken stones, rubble, or burnt ballast should be put in on top of the tiles before filling in the removed earth. With *frequent deep tillage* such drains will work fairly well up to 7 or 8 yards apart, but much will depend on the nature of the clay, for some so-called clay soils are much more open in texture than others.

The cheapest and most thorough method of draining clay land is, in my opinion, a modification of the old high-backed-land system. Set out the field for ploughing in strips or pieces from 11 to 15 yards wide running down the slope of the field. Plough the pieces, commencing in the middle of each strip, so that the soil is heaped towards the centre and furrows are left open every 11 or 15 yards as the case may be. Repeat this three or four times at intervals, taking care to harrow down the crests after ploughing. This may be done gradually

during a year or two if desired, so as not to interfere
with the cultivation of the field, but in the end we shall
have a series of high-backed lands with the furrows in
what was originally the subsoil.   Immediately after the
last ploughing start to put in the drains, making one
run up every furrow.   Two feet will be quite deep
enough in this case; and the pipes, which should be 2
inches in diameter, must be covered with rubble or
ballast, as mentioned previously, before the earth is filled
into the drains.   When finished, the ridges may be
ploughed in the opposite way, to fill in the furrows and
lower the ridges somewhat.   The soil having been moved
once or twice, will be more porous than it was before;
and if we always keep these lands slightly higher in the
middle, the water will naturally tend to run towards the
drain.   This slightly undulating surface need not interfere with any farming operations, and it gives us the
only satisfactory method of thoroughly and cheaply
draining stiff clays.   The cost should run between
£3, 10s. and £5 an acre, including everything except
the ploughing.

Whatever the kind of soil, plan your drains so as to
get a good and sufficient fall.   It is not necessary always
to take the greatest fall, but draining across a hill is
often unsatisfactory.   Where the field is very flat, have
the falls accurately taken with a proper level, even if
you have to employ somebody to do it.   I know a large
field which was drained to run uphill from want of this
precaution, because it appears to fall towards you
wherever you stand.

Now as to leads or main drains: have as few of
them as you can; they are a delusion and a snare.
Carefully avoid the fish's back-bone patterns that look

so pretty on a plan, but work so badly in the field. They are always blocking up; and where it is absolutely necessary to have a number of drains collected into one lead, that main drain should be taken up and cleaned every four or five years. Drains are not going to run for ever without attention, and many a landlord has found them "though lost to sight to memory dear" in more senses than one. A main will still continue to run even if half the sub-drains leading into it are stopped up. Either take every drain straight into the ditch or have main drains of *three- or four-inch* pipes as short as possible. A good way of marking the outfall of the pipes into the ditch is to take two pieces of oak 3 feet long and 3 inches broad by 1 inch thick, and nail them together to form a V-shaped trough. If this be laid in the drain, projecting about 6 inches into the ditch, and the pipes laid in it before filling in, it will mark the place for several years, and the end pipes cannot get knocked out of place.

I have so far spoken only of pipe draining; but on heavy soils, bushes obtained from copses or hedges are often used instead of pipes. Such drains are laid out and dug in exactly the same way as for pipes, and the bushes are then laid along the drain and tightly trampled in, a little straw often being laid on the top before the soil is filled in. These drains are only suited for the heavier soils, and will often last from 10 to 15 years.

The mole drain is another form often used on heavy soils, and although the work can be cheaply done, it cannot be guaranteed to last very long. Its efficacy is greatly increased by having a few pipe drains previously run across the direction of the mole drains at intervals of, say, 100 yards. These pipe drains should be quite 9 inches deeper than the mole is set, and should be filled with 18 inches of ballast or other porous material, so that as the mole passes across them it will not disturb the pipes, and the water which comes down the mole drain will find its way through the porous material into the pipes.

In soils which require it almost any kind of draining pays, but the best pays best. The best draining is not always the most expensive, and when one hears of draining costing £10 to £15 an acre, I am always very sceptical of the judgment which planned it, and of its consequent efficiency. Light soils may and often do require draining, but heavy clayey soils always do ; for them it is the first step towards fertility: it makes the horse labour lighter, cultivation easier, our manure more effective, our harvest earlier, and our crops more abundant. Remember, however, that clay soils are not naturally porous, and that to lay pipes in them is not always to drain them.

Burnt Clay or Ballast is one of the most powerful factors in adding porousness to clay soils. This clay burning is a very old custom, and there are few heavy soils that do not show traces of it in abundant red particles, often to be found in streaks and layers at a considerable depth. So striking are its results very often, that strong manurial properties used to be attributed to it. It is probable, however, that its effects are more due to its mechanical effect on the soil than to any manurial substances it may contain. Great changes take place in the clay during burning, one of the most important of which is that the sticky hydrated silicate of alumina is changed into a porous silicate, incapable of ever again becoming sticky. Some of the insoluble potash is so changed as to become soluble, and any chalk which the clay may contain is changed largely into quicklime, which has a great effect in opening up clayey soils.

I have also found that any fossils and nodules contained in the clay become changed into a fine powder, thus setting free mineral matter not previously available. The red colour of burnt clay is due to the oxidisation of the iron it contains. The change of greatest importance, however, is that it will never become sticky again, but will continue to exist as extremely porous particles of varying size. Now, although 50 tons of sand per acre would make very little difference to a clayey soil, yet a dressing of 15 or 20 tons of burnt clay will make a wonderful difference when thoroughly mixed with the soil. Not only will it increase its porosity and air-holding power, but the larger particles act as scourers for the plough and other implements, which is no mean service, as any one accustomed to tilling clay land knows.

It used to be the custom to pare the surface of pastures and to burn the turf, but this cannot be recommended, as the whole of the nitrogen contained in the turf is driven off into the air during burning. Seeds and creeping roots were effectually destroyed in this way, and there are a few cases where it might still be allowable, but at best it is a wasteful practice. It is better to select a convenient site on the side of a slope where it is possible to get at a good depth of the subsoil, and here start a fire. Small fires produce the best ballast for dressing purposes, being finer, containing fewer hard lumps, and having more soluble potash than that produced in large fires, but large fires are undoubtedly the cheapest and most economical. Hard lumps when produced can be cracked or picked out for road-mending or other purposes, and the coarser material would be better for the drains alluded to.

Ballast fires are easily started with a foundation of wood and coal, and it requires a certain amount of slack coal sprinkled on the clay to keep them going, say half a hundredweight to the cubic yard of clay. For fires of from 100 to 200 cubic yards the cost should be about 1s. a yard.

Liming or Marling is often productive of great benefit to many soils, and the means by which the lime works this benefit are very various. We can apply lime to a soil as quicklime, slaked lime, chalk, limestone or marl; and as gas lime and gypsum, which belong to a different category.

Chalk and limestone consist nearly entirely of carbonate of lime, whereas marl is the term given to a mixture of carbonate of lime and earthy matter in varying proportions. If we can get this carbonate of lime fine enough to mix intimately with the soil, it will produce all the benefits we require. When ground into powder, 2 or 3 tons per acre are sufficient, but when applied as lumps, from 10 to 15 tons are required.

Quicklime is chalk or limestone burnt in a kiln till the carbonic acid is expelled and oxide of lime is left. When water is added to quicklime it falls into an exceedingly fine powder, and becomes slaked lime, which is the most powerful and effective form in which we can apply lime to a soil. Quicklime of course becomes slaked lime as soon as it comes in contact with the water of the soil, and slaked lime eventually becomes carbonate of lime, so that after a time all three forms will have the same action in the soil. Slaked lime, however, is much more active at first, and has a much greater power in neutralising the acidity of the soil than the carbonate, and it also has the power of setting free ammonia from any organic matter with which it may come in contact. If you take a little rotten stable manure and stir into it some slaked lime, you will at once be able to smell the pungent odour of the escaping ammonia. For this reason lime should never be mixed with any kind of compost or dung, although it is a very common practice. I have had my attention called to the strong smell of a heap of compost mixed with lime as a witness of its quality, which is just about as sensible an argument as leaving the tap of a barrel running to show that it has something inside it. When, however, lime is mixed with the soil, any ammonia set free is absorbed and held by the damp earth, and is readily nitrified by the bacteria. Nitrification is always greatly assisted by the presence of lime, not only by the setting free of ammonia, but also because the lime combines with the nitrous and nitric acids formed, keeping the soil sweet and healthy for the bacteria, while it seems probable that the bacteria use the carbonic acid of the carbonate of lime as food.

Peaty soils, especially if resting on clay, are sometimes found to be very infertile, and green and bluish clay subsoils are often quite sterile. The cause is the same in both cases, there being present certain poisonous compounds of iron ; but quick or slaked lime effectually destroys their influence and restores the conditions of fertility. Another very remarkable effect of these forms of lime is the power they have of making clays easier to work and less sticky, by making the very minute particles join together into larger ones. This is prettily shown by a very simple experiment. Shake up some clay with about a pint of water, and when the thick part has settled out, pour an equal quantity of the muddy water into two tumblers. Add a small pinch of slaked lime to one, give a final stir, and leave them for a short time. The one containing the lime will soon be clear, because the particles will be made too big to float in the water, while the other will remain muddy for a long while. Where it can be obtained at a reasonable rate, quicklime is undoubtedly the best form in which to buy lime. It should be carted at once into the field, and put down in little heaps of about a quarter of a hundredweight each ; and each heap should be covered at the same time with a shovelful or two of soil.

In this position the lime will soon become slaked, either by rain or by absorbing moisture from the air, and it can then be spread with shovels. Great care has to be taken in dealing with slaked lime, for, owing to its fineness, it easily rises in clouds of dust, with disastrous effects on the clothes and eyes. For this reason never try to spread lime from carts, but the little heaps enable it to be spread so that the wind will carry the dust away from the worker. From 1 to 2 tons per acre is an ample dressing in most cases, and it should be ploughed in as soon as spread.

Now let us look at the two other forms of lime— gypsum and gas lime. ·Gypsum is a sulphate of lime, and it is capable of exercising none of the beneficial effects of the other forms. It can, however, supply the plant with lime as food, and it is capable of setting free a certain amount of potash in the soil. For this reason it has been found beneficial to clover crops in soils deficient in lime, but the other forms will supply these wants equally well, besides performing their other functions. Gas lime is a waste product of the gasworks, being the lime which has been used to extract the sulphurous compounds from the gas. It therefore consists largely of poisonous sulphides of lime, and rejoices in a most magnificent odour. This otherwise indescribable odour of gas lime must be the seat of nearly all the remarkable virtues ascribed to it by many farmers, for they fail to be found elsewhere.

There is a fascination in the evil odour of any substance which is going to be applied to the soil that seems to exercise a spell over the mind. A manure without a smell must necessarily be useless, and it would fail to gain a reputation, just as a doctor would who omitted to add plenty of bitter extract to his medicines. None of the substances taken from the soil by plants have any smell whatever, but I suppose the connection between manurial virtues and maliferous odours is supplied by thoughts of the farm stables.

The manurial virtues of gas lime, if any, are very infinitesimal, but it always contains a certain percentage of carbonate of lime which no doubt benefits the soil. Fresh gas lime is exceedingly poisonous, and is able to destroy any green plant, fungus, or insect with which it comes in contact. It has been found, however, by experiment, that to destroy wire-worms in the soil, so large a dressing has to be used that not only would the bacteria be destroyed, but the soil would be poisonous to plants for many months at least.

Where gas lime can be obtained for very little more

E

than the cost of carting, it may be used as a source of lime supply, using from 4 to 8 tons per acre. It should, however, either lie in a heap for many months before being used, or it should be carted on to the soil and spread in the autumn and allowed to lie all winter before ploughing in, as it loses its poisonous properties on exposure to the air. Gas lime is often recommended as a preventive of the disease of turnips known as "finger and toe," but as that disease only occurs in soils very deficient in lime, any kind of lime will have the same effect.

It will be seen from all I have said that lime is often a very useful help in farming. Nearly all sandy soils and a few clays, notably that known as the London clay, require lime as a plant food, but on nearly all soils it presents us with a powerful lever wherewith to set free the locked-up stores of Nature. It is this that has given rise to the old saying that "lime makes the father rich but the son poor." Of course in the days when everything was taken off the soil and very little brought back this probably was true, but where larger crops are grown, more stock can be kept and more manure goes back on the land, so that by the aid of purchased foods and artificial manures the land ought to get richer instead of poorer. The real evil lay in the enormous dressings of lime that used to be given in old times, so that much more plant food was set free than could be used by the plants and waste resulted, but science and experience have shown that with lime, as with chemical manures, the rule should be little and often.

Claying.—Although it would take an enormous number of loads of sand to make any appreciable difference to a clay soil, a comparatively small quantity of clay makes a great difference to a sandy soil.

The great difficulty in growing remunerative crops on very sandy soils lies more often in their lack of adhesiveness and water-retaining power than in the absence of plant food. The addition of large quantities of vegetable matter to such soils greatly assists their water-holding power, but they still remain so light and open as to present a very imperfect root-hold to the plants, and soil is even liable to be blown away by very high winds. A gentleman being driven along a Suffolk road asked his driver, "To whom does the soil belong here, my man?" "Well," said the driver, "it depends; for when the wind be in the east it do mostly belong to Mr. Smith, but when it be in the west it belongs to Mr. Brown." I have tramped over many miles of these Suffolk soils, and can well believe in the possible truth of this tale; but wherever clay is available, the farmers periodically dress their fields with it at the rate of from 10 to 20 loads per acre, with excellent effect.

The fen farmers of Cambridgeshire often suffer in a similar way from the high winds, but there, not only is the soil removed, but it is thrown into the dykes or drains, and entails the expense of having it cleared out. Some four or five years ago, after a spell of dry weather, a high wind removed both the soil and the barley which had been drilled, so that farmers I knew had to drill the land again. Here claying is a general practice, but as the clay lies under the peat, long deep trenches are cut like drains at regular intervals, and the clay is thrown out on to the top. The sticky nature of clay makes a small quantity of it beneficial to any soil, for it makes the soil firmer and provides a better root-hold for the plants; it also supplies mineral food, and by filling up the chinks between the larger particles assists the soil to retain moisture.

Green-manuring is, however, the greatest of all factors in increasing the water-holding power of soils, and of light sandy soils in particular. In its strictest sense, green-manuring means growing a crop and ploughing the whole of it into the soil as soon as it has reached its greatest growth. This is a common practice in many parts of the country where the soils are light, and its effects are often little short of marvellous. There are several modifications of green-manuring, some of which are m(re applicable to the heavier soils owing to the danger of large quantities of unrotted vegetable matter causing loss of nitrates by denitrification. This danger need not be feared on light land, but on the heavier soils it is better to plough them in a month or two before sowing another crop, or to let sheep run over the green crop, and thus convert some of it into droppings and trample down the rest, so that it will start decaying before it is ploughed under. We must remember, however, that where the object is to increase the store of humus in the soil, we must not sheep the crop off. That will manure the land, of course, but the sheep droppings will form very little humus.

The best possible crops to grow for enriching the land are the leguminous crops; the most suitable being red clover, trefoil, crimson clover, tares, and lupins.

Red clover may be mown for hay early in June, and the second crop ploughed in during August. Trefoil and crimson clover reach their best by the last week in May, and may be ploughed in and followed with a crop of white turnips on light soils. Tares may be consumed by sheep or ploughed in during June or July, and are particularly suitable for heavy land. Lupins will grow on soils so sandy as to be almost barren, and should be rolled down and ploughed under in August.

All these plants are nitrogen-gatherers; that is, they can obtain nitrogen from the air by means of the bacteria which live in their roots, and therefore when ploughed in they leave the soil richer in nitrogen than they found it. Exactly how much nitrogen these crops obtain from the air and how much from the soil it is impossible to find out. I have seen large crops of lupins growing on a soil so barren that they must have obtained nearly the whole of their nitrogen, probably 200 lb. per acre, from the air, and I think we might safely estimate the captured nitrogen of clover and tares at from 50 to 100 lb. per acre.

Other plants which are not nitrogen-gatherers, but which may be grown for green-manuring, are mustard, buck-wheat, and rape.

A crop sown on land in the autumn merely to use up and prevent the washing away of soluble food by the winter rains, is spoken of in American agriculture as a "cover crop," *i.e.* it covers the ground which would otherwise be bare. These cover crops are of great value, for the formation of nitrates in the soil is often most rapid after our ordinary corn crops have become too ripe to require them. These nitrates would all be washed away by the winter rains, especially on light soils, so that it becomes good practice to sow a quick-growing plant like mustard which will use them up. When mustard is growing on the stubbles, it is usual to leave it till killed by the frost, for when the ground becomes cold no fresh nitrates are made till the spring, and those absorbed by the mustard are safely built into its tissues. The cultivation of the various crops mentioned will be found discussed, each under its own name, in the chapter devoted to farm crops. There is one warning I would give the sporting farmer. Lupins, buck-wheat, and mustard make splendid cover for game, and because of this they are often allowed to stand till past their best as a green-manure crop. They should be ploughed in when green and succulent, and having been caught in the trap myself, I am anxious to warn others : if you want game cover, sow the crops for cover, but if you sow them for ploughing in, plough them in at the right time or you will regret it.

In ploughing in green crops great care must be taken that they are well covered, and this is almost impossible if the crop gets hard and woody ; better a week too soon than a week too late. A roller may be run over the crop in the same direction as the plough is going to travel, and a heavy chain reaching to about the middle of the breast may be hung from the beam of the plough. If properly adjusted it will sweep into the furrow most of the loose tops of the plants.

The principal object of green-manuring, as I have tried to point out, is to supply humus to the soil, so that we may have the benefit of the wonderful assistance it gives us in bringing about fertility; any other effect it may have, such as the gain and prevention of loss of nitrogen, are secondary considerations, though of great value. The consuming of crops on the land by sheep really belongs to manuring, and will be treated of under that head, but at present I am endeavouring to draw attention to the various methods of utilising the "inherent capabilities" of the soil without the direct application of manure.

There is nothing new and original in what I have said; in fact some of the operations which I have recommended were common practice hundreds of years ago, but have fallen into desuetude with the advent of imported feeding-stuffs and chemical manures. The value of cakes and artificial manures cannot be overestimated as means of increasing the resources of the farm; but I contend that before we can obtain the full and proper benefit of any manuring, we must make the soil in a suitable condition for the healthy development of the plants by suitable cultivation; and, having utilised the resources of the soil to the utmost, we can then ensure perfect crops by meeting the deficiencies of the soil and the requirements of the plants with applications of the proper manures. Some land is much more easily kept in the condition suitable for the healthy development of plants than others, and these are the lands which command high rents and are easy to farm. Unfortunately we cannot all get such land, and we have to learn how to make the best of what we have got, and this is why I have taken up so much space with what may be called the more permanent improvement of soils. We have now to consider those operations usually classed together under the name of tillages.

Ploughing is one of the most ancient of arts, but in spite of these thousands of years and its vast importance, the evolution of the plough as ·a tillage implement has been slow in the extreme. When Adam delved he no doubt used a sharpened stick to loosen the surface of the soil, and some inventive grandson was struck with the idea that if he could make an ox draw the stick through the ground he would lighten his own labour immensely. He therefore fixed his stick to a beam on

to which he could fasten his ox, and soon after discovered that a handle behind would enable him to guide
the implement. Thus the plough was no doubt in-
vented, and implements of this description are used in
the East to this day, though now shod with iron.
Ploughs with a coulter and a wedge-shaped breast, and
with the beam supported by wheels, were in use in England nearly a thousand years ago, and some of the
ploughs in use in parts of this country have altered
very little from that day. The idea of the curved breast
which inverts the furrow is a very modern invention,
and with the use of iron in the end of the eighteenth
century the real improvement of the plough began.

In spite of the eulogism of the British plough of to-
day, and of the beautifully finished implements turned
out by our best manufacturers, I am firmly of opinion
that the credit of making the plough an efficient and
economical tillage implement belongs to the manufacturers of Northern America. In making this statement I know that I shall be branded a heretic of the
deepest dye, but I am prepared to stand by what I say.

Let us see then what we want a plough to do. First
it must invert the soil, covering in all stubble, weeds or
manure, and exposing a new surface to the air. In
doing this all deep roots should be cut, the furrow-slice
should be moved so as to leave a furrow up which a
horse can walk, and the moved furrow-slice should be
left in such a condition of tillage that it exposes the
largest possible surface to the action of the atmosphere
and requires the least possible labour subsequently to
fit it for the healthy growth of plants.

The essence of the whole matter is that the plough
should be a tillage implement; it should thoroughly till

the soil and not merely turn a number of very regular and straight slices of soil upside down. About 1730 Jethro Tull endeavoured to make his plough a tillage implement by introducing three extra coulters so as thoroughly to divide the soil, for it had not then been recognised that the breast is the tillage part of the plough.

The judges at ploughing matches are the persons most to blame for the inefficiency of the British plough. To gain a prize at a ploughing match it is necessary to be able to plough the straightest, most regular and even furrow-slices, which must be quite unbroken from one end of the field to the other, and tightly pressed together with a shiny surface and a sharp crest. To such ploughing prizes were awarded, the givers forgetful of the fact that no real tillage had been performed, and that any amount of extra work was necessary before these very pretty-looking furrow-slices could be broken sufficiently to be fit for the growth of plants.

The manufacturer, of course, anxious that his should be the plough with which the prize ploughing was performed, naturally devoted his attention to turning out a plough capable of producing these pretty furrow-slices, with most excellent results. Such ploughs must have long breasts with a very gradual turn, so that the furrow may not be broken while turning over; they must be of considerable weight to give steadiness, of great length to give steering power, and the front must be supported on wheels to give regularity of depth. The English plough is seldom made to take a larger furrow-slice than 6 inches deep and 10 inches wide, but the Scotch plough is made to turn a slice up to 10 or even 12 inches in depth. For the lighter class of soils this

Scotch plough *is* a tillage implement, and we must remember that on clays the depth of the soil will seldom allow us to plough anything like this depth even if we were able to do it, but there is no reason why we should not successfully turn a slice from 13 to 15 inches wide. I have more than once seen a team of three horses in a string, ploughing a furrow-slice 2½ inches deep and 7 inches wide. This sort of thing is an absurd waste of labour, but the reason for it is apparent: a solid unbroken furrow-slice of larger dimensions could never be harrowed into anything resembling a tilth.

What we want is a plough which will produce a tilth as it ploughs, which will thoroughly pulverise the whole of the furrow-slice and leave us a considerable depth in that open and friable condition which is so conducive to a healthy condition of the soil.

Admitting that it is desirable on clays and loams to have the soil thoroughly pulverised to a considerable depth, and that a well-broken and friable surface saves much labour in preparing a seed bed, what kind of plough will produce this? The great essential is a comparatively short and sharply turned breast, which shall not squeeze or slide on the furrow-slice after it is 'urned. With a plough of this pattern the furrow-slice is first sharply turned upwards, and immediately afterwards as sharply turned on one side, so that it is fractured in all directions and falls freely into its place in a thoroughly disintegrated condition. With the long-breasted plough the turning is too gradual to cause any fractures, and the furrow-slice is smoothly pressed into its place by the sliding action of the end of the breast.

It will be seen that the work resulting from these two

types of plough is very different both in effect and in
appearance; indeed, I am convinced that the objection
to the pulverising plough is largely due to the appearance of its work. The farmer has been so used to
associating ploughing with sharp, regular furrow-slices,
that the broken corrugated surface without any distinct
slices looks to him wrong and untidy. Although the
pulverising plough is much the best in every case where
tillage is required, there are times when unbroken slices
can be used with advantage. An unbroken furrow-
slice will dry through more quickly than a broken one,
because for a long time after ploughing there will be
a space between the slice and the subsoil, and water
will not be able to rise from below to keep the
slice moist. This may be an advantage when land is
ploughed up to lie all winter, or when ploughed in the
spring for fallowing purposes and we wish it to dry
rapidly.

The draught of a plough is an important point, for
on the same soil, and with the same sized furrow-
slice, the draught of different ploughs may vary by
nearly 50 per cent., and that which gives the easiest
draught on one soil may not do so on another. I
have also known the same plough with the same sized
furrow-slice, on the same day, to vary in draught from
a little over 800 lb. to a little under 300 lb. on different
fields of the same farm. If a better plough would have
decreased the draught by even 25 per cent., what an
important saving this would be! A draught of 300 lb.
is considered fair work for two horses although they
may take up to nearly 400 lb.; but there are many
sandy soils on which a fair-sized furrow may be turned
with a pull of less than 200 lb. Three horses should

here draw a double-furrowed plough with ease. These double-furrowed ploughs economise much labour, and should be used more than they are.

The steam plough is of great assistance to the farmer of the heavier soils. It enables him to get a large acreage turned over very quickly, which is often of the greatest importance after harvest. There is often much prejudice against the steam plough, but it has largely arisen through injudicious persons ploughing too deeply. Care should be taken to plough only the usual depth with the steam plough, for although a clay subsoil may be well stirred as deeply as we like, it should never be ploughed on to the top. The depth of a soil may soon be increased by frequent stirring of the subsoil and thorough cultivation of the top soil.

Cultivating is a term we have used up to now in the sense of general tillage of the soil, but it is commonly used to denote the operation of stirring the soil with the toothed or tined implement known as a cultivator. These cultivators are of many patterns, being developments of the old-fashioned scuffle, which was a ponderous implement with from six to ten enormous iron tines. Our ideas of cultivators were completely revolutionised by the advent of the Massey-Harris cultivator, a Canadian invention, which sold by thousands immediately on its introduction into this country. The sale of this implement is a standing refutation of the statement that the British farmer will not adopt a new invention. The features of this cultivator are that it is lightly made of steel with flexible tines, it is mounted on high wheels, which give it steadiness in running, and it provides a seat for the driver. Somewhat similar implements can now be obtained from most British manufacturers, and there are also many excellent and light patterns of cultivators with rigid tines. I have seen various kinds of cultivators doing excellent work, and it is difficult to say that any one is the best. For breaking up hard furrows I have heard the rigid tines spoken highly of, but I believe the spring tines to be equally efficacious, indeed their shaking action seems to bring weeds to the surface better. Although cultivators are most frequently used for the preparation of tilths after ploughing, they may frequently be used to prepare a seed bed without ploughing. A neighbour of mine saved a good deal of labour very successfully in this way. A field of wheat stubble, clean except for a few surface weeds, was cultivated over twice, and a beautiful tilth three or four inches deep was produced. In this, winter beans were drilled, and an excellent crop resulted. Immediately after harvest the field was cultivated again twice over, the weeds gathered up, and wheat was then drilled and clover sown in the spring. Both the wheat and the clover were excellent crops, and thus four crops were grown with one ploughing. When the clover was ploughed up the field was quite clean, and produced an excellent crop of winter oats, finishing the rotation. Now this seemed to me a great saving of labour and worthy of trial in these days, for three horses and a man are able to cultivate four or five acres twice over in a day, and leave the land all ready for drilling.

No horse cultivator can be expected to stir the soil deeper than it is usually ploughed, so that when we wish the subsoil to be tilled the steam cultivator must be used. The steam cultivator is a splendid implement on heavy land, and we can break up the subsoil just as deep as we like, provided we plough the land over first.

This is the great secret, for it completely prevents the turning up of huge lumps of subsoil on to the surface, lumps with which no horse implement can deal. I have previously tried to point out the excellent effects of this stirring of the subsoil on the heavier lands, how it deepens and aerates the soil, makes it more porous and permeable by roots, makes it more retentive of useful moisture, and on drained land it allows the surplus water to reach the drains more quickly. This effect will be enhanced if the land has been dressed with burnt clay before cultivating, for it will fall into the tracks of the cultivator tines and tend to keep the subsoil always open. The steam cultivator is unsuitable for light sandy soils, for there the subsoil is better kept solid and the cultivation confined to the top soil.

The typical soil is one half-way between a sand and a clay, and although we try to get a clay more open, we want a sandy soil to become closer to make it more retentive of moisture and to enable the moisture to rise up from below. In the United States it has been found that the size of the crop depends much more on the supply of moisture in the soil than on the actual abundance of plant food. The one is, in fact, useless without the other, and although we can supply plant foods we cannot supply moisture. In our more humid climate the difficulty may not be so great as in America, but our lighter soils, and indeed many of the heavy ones, have suffered much from drought during the last few seasons. We cannot regulate the supply of water *to* the soil, but we can control the moisture *in* the soil to a very great extent, by that which the Americans call *surface tillage.*

The operations of surface tillage are commonly practised in this country, but often without due appreciation of their effects. These operations are harrowing, rolling, and hoeing. Harrowing is usually associated with the fining of the surface of the ground in the preparation of a seed bed, or as a means of getting out weeds; rolling with the consolidation of the soil or the breaking of clods, and hoeing with the destruction of weeds; but they produce other important effects. Water evaporates the most rapidly from solid land, and every farmer has observed the drying effects of a caked surface on a field, as well as the rapid drying of solid furrow-slices and clods. This is due to the rapid flow of water to the exterior by capillarity; and as the moisture evaporates on the surface, more flows along to take its place and is in its turn evaporated, until the whole is dry. It is a common observation again, that when a soil is well harrowed and reduced to a fine tilth the ground becomes more moist underneath, although it may be quite dry on the immediate surface. This is due to the fact that the moisture flowing upwards by capillarity finds its progress arrested by the loose particles on the surface and tends to accumulate underneath them, while these particles act as a blanket to prevent the wind from evaporating the moisture below.

This blanket-like action has also an important effect on the equalisation of temperature, for it is well known that very solid ground is the first to freeze in winter and the last to thaw in spring, and the chilling effects of cold winds are minimised by the warmer air imprisoned among the particles of the blanket. In the very hot sunshine of summer, too, the temperature is kept more even by the screening effects of the finer particles, an effect which we see intensified when the land becomes covered by the crop. In attributing these two opposite effects to this coat of fine soil, I am reminded of the legendary Irishman who is reported to have worn his overcoat in summer because if it kept out the cold it should also keep out the heat. The cases are the same as regards the cold, but in summer the heat of the soil is entirely external, whereas the Irishman's discomfort arose from the heat of his own body.

The effects of surface tillage on moisture are, however, of greater importance than its effects on temperature, and its influence on moisture is of more consequence in summer than in winter.

The excellent effects of hoeing upon crops is quite as much due to the loosening of the surface and consequent retention of moisture as to the killing of weeds. We cannot horse-hoe and otherwise cultivate between the rows of potatoes and turnips too much, provided we do not till so deeply as to injure the roots. In corn crops, where hoeing is too expensive or inexpedient, we can produce a similar effect by harrowing the surface. If a light harrow or an American "weeder" be used when the soil is fairly dry, no injury to the young plants need be feared. Wheat is often greatly benefited by harrowing in the spring, even although it may be four or five inches high. Rolling is commonly practised to consolidate the soil, but where this is done great loss of moisture will result, unless the surface be lightly harrowed soon after. In the case of turnips, clover, and other small seeds which can only be lightly buried, it is common to roll after sowing, because the soil will become moist up to the surface and cause the seeds to germinate, and the moisture has to be sacrificed to the end in view. It has been found, as an average of many experiments, that although the top 18 inches of soil gained by rolling to the extent of 6 tons of water per acre, the next 18 inches had lost more than twice as much by the upward movement and consequent evaporation.

Harrowing, to be effectual, should not merely scratch the surface, but should thoroughly pulverise the top two or three inches. I have found the spring-toothed chain harrows and weeders much more effective in this respect than the ordinary straight-toothed harrows. I also prefer the ribbed or Cambridge roll to the smooth roll for compressing purposes, as the surface left is corrugated and not smooth.

The last of the tillage operations which I shall touch upon is fallowing. The term implies leaving the land without a crop, but this is not and should not be always the case. The idea of resting the land is a fallacious one; land never requires a rest, but it may require cultural operations which prevent it being cropped. Fallowing upon light and heavy soils are two entirely different operations, and I have known the failure to recognise this cause a good deal of trouble. Upon the lighter soils fallowing merely means cleaning the land of couch grass and other perennial weeds, and such cleaning can be accomplished in time to sow a crop of swedes or turnips. The cleaning may begin in the autumn by means of the horse cultivator, or the land may be ploughed during the winter, and as soon as it is dry enough the cultivator and harrows may be used to pull the weeds to the top, so that they may die or be removed. The land may then be ploughed again, and the cultivating and harrowing repeated until it is clean. Care must be taken to plough deeply enough to get below the runners of the couch grass,

F

and when clean the land may be manured and prepared for the root crop.

In heavy land, however, the couch grass and other weeds cannot be pulled out of the soil on to the top, but must be killed in the soil by drying in the sun. Even if it were possible occasionally to pull out the weeds from a clay soil, we should not derive all the benefits of the deep, thorough cultivation and aeration of the soil which is so productive of good results in this class of land. An occasional dead or whole summer fallow gives us the opportunity of draining, steam-cultivating, and thoroughly shaking up the subsoil, besides cleaning the land of weeds. To obtain the best results by dead fallowing the land should not be ploughed till it has become dry in the spring, the end of April or beginning of May being soon enough to plough. When the land is thoroughly dry it may be ploughed again across, but it is usually a mistake to harrow or cultivate till most of the weeds are dead. When this occurs, say by the end of June, the steam cultivator may be used thoroughly to stir the subsoil, after which little more should be required till the final ploughing in August.

Half-fallows or bastard-fallows should be obtained wherever possible ; that is, the operations of a fallow should be performed after the removal of such a crop as clover or tares. Any class of land may be kept fairly clean for a number of years by the occasional inter-vention of such a fallow.

The increased fertility of heavy land after a fallow has no doubt given rise to the idea that a rest was necessary ; in fact I have often heard it said that a fallow is as good as a coat of muck ; but I think it will be seen that this increase is due to the thorough cultivation. The effect of the thorough aeration upon the decay of vegetable matter must be great, and the nitrification very rapid. It is owing to this rapid nitrification and the consequent fear of the nitrates being washed away by rain that dead fallows have been so frequently condemned by scientific men. On light soils a dead fallow would of course be absurd, but my experience is that clays require it periodically.

CHAPTER IV

Manuring

I HAVE tried up to now to show what are the natural capabilities of the soil with regard to fertility, and how its fertility may be increased, and those natural capabilities made useful to us. It may be thought that I have spent too much time on the soil and its cultivation, but I am firmly convinced that the tendency of the present day is to rely too much on manures, and too little on thorough cultivation and the other means of improving the fertility of the land which we have on the farm itself. On the light sands and loams I know the cry is "manure, still manure," and the cultivation means simply keeping the land clean, but even there much might be saved by growing green crops for ploughing in or for sheep feeding and to prevent the washing away of nitrates. The natural store of plant food in light soils is always less than in those of closer texture, because the conditions of porosity and air supply encourage the growth of bacteria and the making

available of the organic matter; in fact they often
suffer from an excess of the very conditions we so
desire to bring about in the heavier lands. In light
land the plant food tends to become soluble and available more rapidly than it can be used by the plants,
and hence often gets washed away, while in the heavier
lands the difficulty is to get the plant food to become
available at all. This is the case not only with the
natural stores, but also with the plant food supplied as
manure.

The following instances, taken from two experiments with which I was personally connected, well
illustrate this undoubted fact. A dressing of 10 tons
per acre of good farmyard manure was ploughed in
for swedes, and, without further manuring, barley
was grown in the second year, clover and rye-grass
hay in the third, and wheat in the fourth. Compared with a portion receiving no manure at all, the
increases given by the 10 tons of dung were, on the
light sandy land, 7 tons of swedes, 6 bushels of barley, 13¼ cwts. of hay, and 2 bushels of wheat, or, at
market prices, an increased value of £6, 6s. due to the
manure. On the heavy undrained clay soil the same
dressing in the same years produced increases of 1 ton
8 cwts. of swedes, 5½ bushels of barley, 2¾ cwts. of hay,
and no increase in the wheat, giving an increased value
of only £2, 2s. 7d. for the use of the same quantity
of manure. When, however, soluble readily available
manures were used, the positions were reversed; 14
cwts. of chemical manure giving a return during the
four years of £6, 18s. on the heavy land, and only
£5, 16s. on the light. These results well illustrate
what I may call the essence of manuring; to manure

successfully, we must apply plant foods to the soil in such a state that they are readily available to the plant.

Even poor land may contain, as I have already pointed out, large quantities of plant food, but the addition of a small quantity of some constituent in an available condition may double the yield. The constituents that are usually most deficient in available supply are nitrogen, phosphoric acid, and potash; some soils being deficient in one, some in another, some in all three, while it is also an undoubted fact that some plants experience more difficulty in obtaining one particular ingredient than another. It is well known, for example, that swedes are benefited by phosphates and seldom require nitrogen, while wheat responds more to nitrogen than anything else. It is in meeting these peculiarities of the particular soils or the requirements of certain plants that the chemical manures are so exceedingly useful.

I am a great believer, however, in fully utilising the resources of the farm, and it is only where these resources are insufficient or require assistance that I advocate the use of artificial manures; but to use them to ensure a good crop is nearly always true economy.

The residues from stock-keeping are the principal sources of manure on a farm, the manure of the farm-yard, and the manure of the sheep-pen; but there are also the residues of crops, such as the roots of beans and clover, and the ploughing in of green crops. A ton of farmyard manure contains from 10 lb. to 15 lb. of nitrogen, about 12 lb. of potash, and from 5 lb. to 8 lb. of phosphoric acid, but its value depends on

FROM THE RIVER PASTURES

the availability of these substances. The straw and the solid excrement contain about half the nitrogen, a large quantity of the potash, and all the phosphates, but these substances are all unavailable to plants until decomposed by the action of bacteria in the presence of moisture and air. The urine, however, contains some of the potash and the other half of the nitrogen, which is in a soluble condition, easily nitrified and made available to the plants. As soluble nitrogen compounds are the most valuable constituent of any manure, the urine is the most important part of farmyard manure. The amount of nitrogen in the manure, and particularly the quantity of soluble nitrogen, as urea and similar compounds, depend to some extent on what the animals are eating, and the kind of animal consuming the food.

The animal takes out of the food what it requires for its own body, and although its principal requirements are starch and sugar (carbohydrates), which are of no value as manure, it also wants nitrogen, potash, and phosphates to build up its muscles and bones. Growing animals and milking cows use up a considerable quantity of these substances in their bodies and milk, but full-grown fattening animals retain very little indeed, so that about 90 per cent. of the nitrogen in their food is returned in the manure. When we consider that the soluble nitrogen in the urine of a fattening beast, in a single day, would cost us from 1½d. to 2d. to buy as manure, we are all the more astonished at seeing this valuable substance being allowed to be washed away down the drains, as is so commonly the case. The value of yard manure as a rapid and efficient fertiliser entirely depends upon the care we have taken of the urine, and manure washed with rain is of very little more value than so much rotten straw.

In my opinion yard manure should always be thrown into a heap to ferment and partially decay before being ploughed in, particularly on heavy soils. This is an old-fashioned plan, which, I believe, has become less common owing to fear of loss of ammonia, but it is the best nevertheless.

The loss of ammonia from a damp heap of manure is not great at any time, but one can prevent it by sprinkling a few handfuls of superphosphate with each load of manure, or by coating the heap with a little damp earth. The advantages are greater availability of the constituents owing to the rotting; the prevention of any possible loss of nitrates being caused by the presence of unoxidised organic matter; the killing of all the weed seeds it may contain; and the greater ease with which it is ploughed under the soil.

Farmyard manure produces a good deal of its effect by the humus it adds to the soil, and it has been found repeatedly in experiments that it pays better to use comparatively small dressings, and to supplement them, if necessary, with dressings of a chemical manure, supplying any ingredient the soil or crop may particularly require.

Upon the same farm, in the same year, 12 loads of dung produced 15 tons 2 cwts. of swedes per acre, but 6 loads of dung and 4 cwts. of superphosphate produced 18 tons 14 cwts., because the dung supplied sufficient nitrogen, but not enough phosphate for the requirements of the swedes. In another case, 20 loads of dung produced 23 tons of mangolds per acre, but 10 loads of dung and 2 cwts. of nitrate of soda produced 27 tons, because mangolds require more nitrogen than dung supplies in an available form. With dung, as indeed with any other manure, there is soon reached a point over which increased dressings fail to pay for themselves. If 10 tons of dung give a certain increase, 20 tons will not double that increase, and 40 tons would probably produce very little more than 20. This is well shown by the following figures, giving the average produce of swedes on nine different farms in the same season :—

	Tons.	Cwts.
With no manure	6	0
With 12 tons of dung . . .	17	12
With 18 „ . . .	19	8

Numerous other striking examples could be given, but it has been often proved that small dressings, supplemented with artificial manures supplying the particular ingredients the soil or crop may specially require, give the most economical method of using this important by-product of the farm.

Sheep are, however, the best manure-cart after all, and the "golden hoof," as it used to be called, still retains its fertilising effect, although the corn crops produced are less valuable than they used to be. In the old days it was considered enough for the sheep to pay for their cake, but I believe that now, in spite of the enormous imports of mutton, it is possible to keep a large head of sheep at a profit, and have their fertilising effects into the bargain.

When a crop is consumed on the land by sheep, from 75 per cent. to 90 per cent. of the nitrogen contained in the crop is returned to the soil, much of it in a readily available condition. Some of the potash is retained in the body and wool of the sheep, but the demand upon the phosphate of lime for the formation of muscle and bone is very much greater. It is therefore necessary on most soils to see that the supply of lime and phosphate is kept up, and this is most easily done by manuring the crops we intend to be consumed by sheep with a phosphatic manure such as superphosphate of lime or basic slag. When we consider that the greater part of the phosphate used by plants is stored in the seed which is usually sold off the farm, and that for generations the animals sold have been carrying off phosphate in their bones, it is not surprising that applications of phosphatic manures should be essential in many cases.

I know no more economical and effectual method of manuring land than to grow a leguminous crop manured with phosphates, and then to consume it on the land. First of all, a dressing of 3 or 4 cwts. per acre of superphosphate will encourage larger and more nutritious crops of red clover, crimson clover, trefoil, or tares; and secondly, the greater the growth of the crop, the more nitrogen it will capture from the air. Now when the crop is fed off, not only have we a large quantity of captured nitrogen put into the soil in a readily available form, but we have the residue of our phosphates as well. In the case of turnips, cabbages, kale, rape, or mustard, we have no gain of nitrogen to the land, although much of this substance which the plants took out of the soil is again returned to it in the available form. Even the use of large quantities of cake will not make the consumption of these crops enrich the land to the same extent as with tares or clover. The value of consuming crops with sheep and of using cakes is principally due to the supply of available nitrogen left in the top soil, but we must remember that this may not be all that our soil requires.

Much waste and loss often arises through manures being unable to exercise their full effect, for want of balance in the plant foods they supply, or because the soil lacks some necessary ingredient. The addition of the particular ingredient wanting in the form of a chemical manure often produces marvellous effects. The land I have been farming in Cambridgeshire is so deficient in available phosphates that a nitrogenous manure merely gives a small increase in the amount of straw without any increase in corn, but the addition of 2 cwts. of superphosphate has increased the yield by from 8 to 16 bushels.

A thin, chalky soil in Norfolk had been found very difficult to farm successfully, but experiments showed that it required potash. In one case 2 cwts. of muriate of potash gave an increase of 33 bushels in the yield of barley, and now every farmer in that neighbourhood uses potash in his manures. Soluble nitrogenous manures give good results on nearly all soils, especially if the other ingredients are present in the soil in fair quantity.

From this we have been able to gather that the effect of a manure depends on several factors : the presence of certain soluble constituents in the manure itself, the natural composition and previous treatment of the soil, the peculiar requirements of particular crops ; and we may add another—the influence of the weather.

Clays, as a rule, contain abundance of potash, large quantities of stored-up nitrogen, and enough lime for the purposes of plant food ; but the nitrogen is very slowly available, phosphates are very deficient, and lime is often required for its action on the soil itself. Chalky soils are usually deficient in nitrogen and potash, but contain a fair supply of phosphates. Sandy soils tend to become deficient in all the constituents of plant food, but particularly in nitrogen, potash, and lime. Loamy soils, on the other hand, being mixtures of all the soil constituents, should contain a fairly well-balanced supply of plant foods, but much will depend whether they incline to be clayey or sandy. Black or peaty soils contain abundance of stored nitrogen, and although they are usually deficient in lime, much depends on the nature of their subsoil, for the soil partakes of the same nature.

The special requirements of each of the farm crops will be given in the part devoted to their treatment, and with regard to the weather it has been found that dry seasons are particularly unfavourable to the formation of nitrates in the soil, and in fact to the utilisation of all manures not in a soluble condition. Moist, warm summers result in the rapid formation of nitrates, and consequently in large strawy crops, while wet autumns and winters cause great waste of nitrates, potash, and lime by washing them out of the soil. It has been found at Rothamsted, as the average of twenty years, that the loss of nitrates by drainage from unmanured, uncropped land is equal to that supplied by a dressing of 2 cwts. of nitrate of soda per acre per annum. Phosphates are not washed out of the soil to any appreciable extent, and by judicious cropping we can, very largely, prevent this astonishing waste of the other substances.

The purchase of manurial substances containing

nitrogen, phosphoric acid, and potash has now become so important in modern farming that it will not be out of place, I think, to consider some of the forms in which those substances can be best obtained.

For the supply of nitrogen our principal sources are nitrate of soda and sulphate of ammonia, and these are the two most readily available and most concentrated forms in which nitrogen can be bought. Nitrate of soda is available to the plant immediately it is applied, and, being dissolved by the moisture of the air, becomes available in the driest seasons; but in wet seasons, on porous soils, there is some danger of it being washed away by heavy rains. It should never be applied till the spring when the crop is well rooted, and then this danger is very slight. Sulphate of ammonia requires to be nitrified in the soil before it becomes available for plant use, and is therefore slower in its action than the nitrate; it requires a moist season to give its full effects, and until it is nitrified it cannot be washed out of the soil. It can therefore be applied to the lightest soils and to winter crops without fear of loss. Nitrate of soda should contain not less than 15 per cent. of nitrogen, and sulphate of ammonia 20 per cent., so that the manurial value of 1 cwt. of nitrate of soda is equal to ¾ cwt. of sulphate of ammonia.

From several years' experience of these manures I am convinced that neither their percentage of nitrogen nor their price per ton can be taken as a guide to the difference in their practical value, for nitrate of soda is by far the most effective on heavy soils for all crops, sulphate of ammonia rather the better for the lighter soils; while on medium loams the nitrate would be better for wheat, oats, and mangolds, and the ammonia for barley, turnips, and potatoes.

Soot is a very popular fertiliser, owing its manurial value to the presence of sulphate of ammonia, of which a good sample will contain about 3 per cent. It contains no other fertilising ingredient, the black colour being due to carbon, which is of no value as a manure and is quite indestructible in the soil. Its fictitious money value seems due to the fact that the carbon can be seen in the soil for many years, and that its general nastiness gives it the proper characteristics of a manure. Sulphate of ammonia being a clean, white powder, could not, of course, possess the same virtues.

The nitrogenous guanos of former days were excellent, effectual manures, but practically none are now obtainable, while the modern guanos of a nitrogenous character contain most of their nitrogen in an insoluble form, and are seldom worth the price demanded. Blood manures, fish guanos, and rape cake all contain considerable percentages of nitrogen, but owing to the slowly available state of this nitrogen they are only suitable for the lighter classes of land, and their price is usually much in excess of their real value. The unwarrantable prices which some of these manures command is due to the prejudice which exists in favour of manures of animal origin, although it has been proved over and over again by experiments that the origin makes no difference to the crops grown. I am reminded of a true story of a farmer whose enthusiasm for this class of manure caused him a good deal of trouble. This farmer was approached in the market by a dealer who offered to sell him several trucks of salmon for manure. The farmer had bought sprats for a similar

purpose, and so a bargain was struck for two truck-
loads. The salmon duly arrived at the station, but
imagine the farmer's horror when he found that his
salmon was all done up in 1 lb. tins! Not to be
beaten, the tins were thrown about a field, and an
attempt made to liberate their contents by chopping
them with bill-hooks. Words fail altogether to express
the effects of the explosion of the badly blown tins,
and the attempt was abandoned. Finally, after a good
deal of trouble with the men, a heavy roller was run
over the field, and the dreadful tins were exploded
and crushed or squeezed into the ground whole. I
never heard of the effect of this manure on the crop,
but the effect on the farmer was such that there was
considerable personal danger to any one who mentioned
salmon in his presence.

Phosphatic manures have been used for a longer
period than perhaps any other manure not produced
on the farm. Bones crushed into half-inch pieces were
largely used for turnip-growing in Norfolk and Suffolk
early in the nineteenth century. It was soon found,
however, that the finer these bones were crushed or
ground the better were the results, and bone meal and
bone ash came into use. In many parts of Cambridgeshire the old-time farmers used to send their carts
miles for road-scrapings owing to their manurial effects,
though I dare say they little suspected the cause. I
have found that the roads were principally repaired
with stones picked up in the fields, and that these
stones in many districts contained large numbers of
the phosphatic nodules which were afterwards known
as the Cambridgeshire Coprolites. On the discovery
of the effects of treating phosphates with acid, these

coprolites became a valuable asset, their digging employing a great deal of labour and bringing much money to the landowners. The supply was, however, soon exhausted, and it is a strange irony of fate that the farmers on these very estates have now to buy large quantities of phosphates imported from Belgium, Algiers, and America.

The value of a phosphatic manure depends altogether upon the amount of phosphate it contains which is soluble either in water or in the very weak acid of plant roots. The analysis of a soil with strong acid shows that an acre of ordinary soil 9 inches deep may contain 6000 lb. or more of phosphoric acid, or as much as is contained in 8 or 10 tons of bones. Only a small quantity of the phosphoric acid of the soil, from 50 to 200 lb. per acre, is soluble by the roots at any one time, and hence the difficulty plants have in obtaining sufficient phosphorus for their requirements. What then is the use of adding a few hundredweights of insoluble phosphate to this already enormous store ?

It is the same story of *availability* as in the case of nitrogen, and there is no doubt but that on light soils more phosphates become available in proportion to the total store than in the case of heavy soils. It is only upon these lightest soils that ground bones, bone meals, or phosphatic guanos may be expected to produce any immediate results.

The price of bones is, in my opinion, prohibitive; but I have seen phosphatic guanos containing nearly 60 per cent. of phosphate of lime at about £5, 5s. a ton, which were probably cheap for use on light land. Dissolved bones are an excellent manure, containing a large percentage of soluble phosphate and some nitrogen, but the price is far too high.

Superphosphate of lime and basic slag are the best and cheapest forms of phosphates obtainable at the present time. Superphosphates may be obtained containing from 26 per cent. to 35 per cent. of phosphate of lime made soluble, and they are suitable for every purpose for which phosphates are applied, and upon any kind of soil. Basic slag has, however, been found to be slightly better than superphosphate for grass upon heavy soils, but I do not remember having seen a case in which it proved superior upon the ploughed land, even of the same kind. Basic slag contains some free lime, and is strongly alkaline, and may possibly have some effect in assisting nitrification, while superphosphate is always slightly acid. A manure called basic superphosphate has lately been put on the market, and as it contains its phosphate in the form soluble by root acid, and is alkaline in its action, it may be advantageous for application to soils inclined to be sour.

There are practically no soils that are not benefited by occasional applications of phosphates, for an abundance of available phosphate is usually the factor that determines the yield of corn. The nitrogen may be said to determine the growth of the crop, the amount of leaves and straw ; but phosphates, and very occasionally potash, seem to produce proper maturity and ripening, and determine how much corn there will be in proportion to the straw.

I have repeatedly seen experimental plots dressed with phosphates ready to cut ten days before those unmanured. The effects of phosphates on heavy land,

G

and potash on light land in giving strength to the straw, are also very remarkable, it being not at all an uncommon thing in experiments to see all the plots badly laid, except those which have received one or both of these manures.

The lasting effects of phosphates are very pronounced, especially of course on land badly in need of them. On my own farm the most casual observer could not fail to see the shapes of plots which received 5 cwts. of superphosphate six years ago, and are now sown down to permanent pasture.

Potash can be obtained in the form of muriate of potash, sulphate of potash, and as kainite. Here there is no question of availability, for all three are completely soluble in water. The muriate and the sulphate each contain about 50 per cent. of potash, and although it is sometimes said that one is better than the other for a particular purpose, I have never yet seen one prove its superiority over the other.

Kainite is a natural mixture of many substances, and contains about 12 per cent. of potash, about 30 per cent. of common salt, besides magnesium sulphate (Epsom salts), and other substances. Kainite therefore contains only about one-fourth of the potash contained by the other two, and when applied in proportionate quantities, there is little to choose between them, except that where common salt may be of benefit to the soil or the crop, kainite may have the preference. As all three substances come from the Government mines at Stassfurt, the prices are always proportionate.

Potassic manures must be used with great care, for where they are not required they often produce considerable decreases in the crop. The cause of this is still unexplained, and is the more remarkable when we consider the strikingly beneficial effects they produce on soils which require them. Mangolds and potatoes are nearly always benefited by potash, and a very small dressing often considerably improves the colour of barley, while the feeding quality of grass is often greatly improved by such small dressings.

This then is a brief outline of the properties and uses of some of the principal artificial manures, and although every one may not agree with me, these are the conclusions I have arrived at, not only from my own experiments and experience, but from others in many parts of the country that have come to my knowledge. To arrive at an accurate understanding as to the requirements of his own farm, every farmer should be an experimenter, and the results of his experiments should always be expressed in accurate figures, arrived at by measuring the ground and weighing the produce. To say "the crops produced have been excellent," or, " I have had splendid results from using such and such a manure," means nothing at all. The business man wants to know how much was the crop increased by the manure, what did it cost, and is it not possible that the same result might have been produced at a less cost ? It is for these reasons that a farmer should always test the effects of any particular treatment, by marking out a quarter or a tenth of an acre to be left untreated. Mark out, at the same time, a plot of the same size on the treated portion, and have the two pieces carefully thrashed or weighed separately. The results can then be expressed in £ s. d., and are of some value. It is of no use to attempt to judge the results by the eye, for I have many times found that with corn it was not always the largest-looking crop which gave the largest yield.

There are large numbers of mixed manures on the market, barley manures, mangold manures, and many others, but although some of them are excellent manures and very suitable for their purpose, they are nearly always much too dear. It is much better to buy the ingredients separately from a well-known and reliable firm.

These chemical or artificial manures are of the greatest possible assistance in supplementing the natural supplies of the farm, and do much to increase its general productiveness. When they are used with judgment, there is no fear of them robbing or spoiling the condition of the land. The nonsense that has been talked of nitrate of soda in this respect shows a great want of knowledge; in fact it is my experience that those who condemn chemical manures most strongly are either those who have never used them at all, or have used them with a great want of judgment.

The best and most successful farmers are those who carefully utilise every method which will enable them to grow the largest possible crop on every acre. To grow the largest possible crop on every acre seems to me to be the acme of good farming, for it means abundance of corn and food for stock, and consequently a probable profit. A mistaken notion of the principles of fertility, and fictitious values put upon the manurial residues of feeding-stuffs, have led to the common belief that good farming consists of bringing on to the farm large quantities of foreign corn and cakes. There never was a more mistaken nor indeed a more suicidal policy for farmers to pursue.

We must buy feeding-stuffs and we must buy manures when occasion requires, but we should regard them as necessary evils; indeed, my ideal of good farming is not gauged by the farmer's purchases, but rather by the large quantity of corn and stock he is able to raise and sell with the least possible extraneous expenditure. This sounds like the rankest possible heresy, but it is true nevertheless, for you cannot produce good crops without good farming.

I remember once walking over a 1000 acre farm with a party of agricultural students, and we were all struck with the quantity of magnificent stock and the luxuriance of the crops. One of the students, anxious to gain some information as to the courses of cropping followed, asked the farmer upon what system he farmed. "I don't know that I farm upon any system," said the farmer, "unless it is this: *sell all you can and buy nothing.*" Observing the look of astonishment on the student's face, the farmer hastened to add, "Well, buy nothing unless you are very sure it's going to pay." I need hardly add that this farmer came from north o' the Tweed, but this characteristic remark sums up the essence of business, and more particularly the spirit of successful farming.

That this man made his farming pay was well known ; and that he translated his maxim in the spirit and not in the letter was shown when we found out his expenditure. Farming suitable land with easy access to a large town, his principal products were milk, hay, and potatoes, and a rough calculation showed that his expenditure on manure averaged fully £1 per acre per annum, besides a fair amount of feeding-stuffs for his 150 cows.

On another occasion I remember a farmer pointing
to an enormous heap of manure with the remark,
"There are several hundred pounds' worth of cake in
that heap." Now the crops on this farm were a very
ordinary-looking example of the four-course rotation,
and the bullocks fed on that cake might have paid or
might not, but I could not help thinking how much
better it would have been had he pointed to a flock of
sheep and said, "There are several hundred pounds'
worth of sheep in that field, all raised entirely on the
produce of the farm." The profit would have been
very much more apparent in that case, for he would
have had something to sell.

In the course of my wanderings up and down the
country I have been over a considerable number of
farms, good, bad, and indifferent. I have been shown
the results of lavish expenditure, magnificent buildings,
thousand-guinea horses and bulls, and luxuriant crops;
I have seen farms where the whole of the work has
been done by the farmer and his family, and where
the expenditure has been little beyond the rent and
taxes; and I have seen many between these extremes;
but wherever financial success was apparent, I have
noticed the same common characteristics. The first
is that there have always been plenty of products of
different kinds to sell, and the second, that the expenditure in producing them has been proportionate to the
probable returns from those products. This seems
simple enough, but it is not always observed. Farmers
do not, as a rule, produce nearly as many saleable products as they might, and this is largely due to the
erroneous notions prevalent as to what constitutes
good farming. To fatten a certain number of sheep

or bullocks each year on foreign food - stuffs does little either to increase the wealth of the country or the wealth of the farmer, but to produce a larger quantity of saleable products by means of increased crops grown on the farm will increase the wealth of both.

The farmer, as an individual, has to be exceedingly careful as to his expenditure, but carefulness is far removed from parsimony. Last summer, when I was walking through an extensive hop-garden, the grower told me that his crop would cost him £60 an acre before it was ready for the market, but that he was determined to grow as large a crop as possible, as there was every prospect of good prices. Judging by the crop and by the prices hops realised last autumn, his energy was well rewarded, for his yield must have been considerably over £100 an acre, and quite three times that of some of his neighbours.

A thousand-guinea bull is a good investment if we are sure of a market for hundred-guinea calves. An expenditure of £10 or even £15 is well justified on an acre of potatoes or cauliflowers, but how much can we afford to spend on wheat, barley, oats, or turnips? The farmer has to be constantly asking himself such questions. Will it pay the better to spend £10 on a crop of potatoes, or £4 on a crop of wheat, or £3 on clover or tares to be consumed by sheep? Or, again, will it pay to spend an extra £3 on labour and manure for the potatoes, or £1 on the wheat, and will it pay to buy some food for the sheep to consume with the clover or tares? In all these things the farmer must carefully exercise his own judgment, but it must be remembered that there are both extravagant and economical methods of bringing about the same result. It does not pay to farm our land badly or to starve our stock, but the most profit comes to the man whose skill and knowledge enables him to produce the best results at the least possible cost.

The difficulty of obtaining the labour necessary for carrying out the operations of the farm has been a very important question of late. In this we are suffering, I believe, for the errors of our fathers, but with our assistance it will work its own cure.

The farmer of the past has always grudged the labourer his wages; he has beaten him down to the last penny, and he would pay him anything rather than money. Payment in kind has practically disappeared, except for the disgraceful and pernicious custom of beer-giving, to which both men and masters still cling in many districts, much to the prejudice of both. I have found it much cheaper to myself, and more satisfactory to the men, to pay from 24s. to 30s. a year in a lump sum in lieu of beer.

The extra payment for harvest is another relic of the days of miserable and insufficient wages, and of times before machines made harvest less tedious than hay-time. On a well-managed farm no extra labour should be required at harvest, for machines now do most of the work with great rapidity.

The early morning work is no longer required; our binders and grass mowers cannot work in the dewy cool of the morning, so important to the wielders of the scythe or sickle, and we no longer carry our wheat to the stack with the dew on it. In hay-time and harvest it is the evenings that are most valuable, and well-paid overtime then should be required.

We in the eastern counties are supposed to pay the lowest wages in Great Britain, but I have calculated that the pay and allowances of my men average 17s. a week without piece-work or Sundays, a wage which, when we consider that a cottage and a rood of land can be obtained for 2s. a week, compares favourably with 28s. or more in a town. The regular wages are, however, only 13s. or 14s., and it is my contention that to make yearly agreements, to pay higher weekly wages, without allowances, and to have a proper scale for payment of overtime, would lead to more contentment.

Good skilled labourers are extremely scarce, and receive higher wages; but no effort is ever made to teach lads to become hedge-cutters, thatchers, trussers, milkmen, or shepherds. No opportunity is ever given for young fellows to acquire skill, and therefore to command increased pay, and I am firmly convinced that the inability to "get on" is the cause of the best young men migrating to the towns. "Hope springs eternal in the human breast," but what hope can spur the efforts of the young labourer on the farm? He receives full wages at eighteen, and he knows that he can get no more; he is liable to be sacked in any slack time, his wages get less as he gets older, and the workhouse is his probable end. In the town, though he start at 18s. a week, there is no limit to his advancement but the ceasing of his own efforts.

The flow to the towns cannot always go on, and we can do something to make matters more equal. The skilled and trustworthy men are worth far more to us than the loafers. Can we not encourage such men by constant employment and somewhat higher wages? I am sure it is worth every farmer's while to encourage brighter, quicker, and more intelligent work among his men, for I believe it will pay well for encouragement in every way. Higher weekly wages seem bound to come, but I am convinced that this country will before long realise that agriculture is an industry of national importance, and with the encouragement of better education, better methods, and better conditions all round, it will be possible to employ more capital per acre with a prospect of profit, and to pay higher wages with more ease than now.

It is truly wonderful how machinery has enabled us to meet the decreasing supply of labour; and the energetic farmer has so thoroughly availed himself of every invention that his labour bill is very much less than twenty, or even ten, years ago. There are still times on the farm when a rush of work has to be done, but modern appliances enable us to cope with these busy periods with comparative ease, and without the employment of extra labour.

The great bustle and excitement of harvest are now rapidly disappearing under the influence of the sheaf-binder and largely reduced areas. Its picturesque side will soon only be preserved to us in the paintings and poems of the past, but fortunately nothing can rob us of the beauty of the fields of ripening corn, nor of the autumnal glory of their surroundings.

In these days of stock-keeping hay-time often rivals harvest in importance, but the mowing-machine has replaced the scythe, the tedding-machine and the swath-turner have taken the places of the fork and the rake, so that the merry gangs of haymakers are but seldom seen in our fields.

Is farming any the less interesting on account of these

changes? In my opinion the interest is increased on every point. The enlarged scope of our resources, the greater certainty of the result of our operations, and the lessened reliance on the caprices of the weather all tend to increase our interest in farming, both as a business and as a pleasure. The gardener who contemplates his roses and well-clipped lawn can afford to ignore the business side of his operations; but is his pleasure any the greater? I doubt it; for I am sure the crown of all his exertions would be unrealised if no dear friend ever trod his lawn or accepted a gift of his roses. The farmer, too, can add to the pleasures of his friends by his farm, but his greater products can only be realised in the markets.

The farmer who makes every possible use of the discoveries of science or the inventions of the engineer will find the greater pleasure in the increased results he obtains for his knowledge and skill. Will the new-turned furrow smell any less sweet because we are using the latest invention in ploughs? Will the lark spring any less blithely from the dewy meadow because the grass grows the better for use of basic slag, or the waving corn rustle any less joyously because of the nitrate of soda? Never! The earth will bring forth her increase all the more cheerfully because we are following Nature's paths.

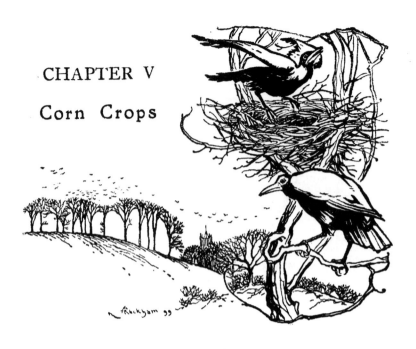

CHAPTER V

Corn Crops

CONSIDERING the means at our disposal for improving our land by thorough cultivation and proper manuring, I am convinced that we might produce larger crops and keep far more stock than we do. It has been repeatedly urged that the keeping of live-stock is the sheet-anchor of the British farmer, but to keep a large head of live-stock it is essential to grow abundant crops of many kinds. The sowing down of pastures in a wholesale manner has been one of the great mistakes of the last few years. The sheep farmer knows that his sheep pay best when kept on the ploughed land, and the prosperity of the Danish farmer is due to the whole of his cows being kept on the produce of land under the plough. We should, I am sure, aim at the greater production of food for stock on our ploughed land; and I may safely assert that without our breaking up any of the best pastures, quite twice the stock could be kept in this country without in the least diminishing the present yield of corn. The growth of forage crops and the consequent increase in the supply of manure would give us the means of producing more abundant crops of grain, so that the value of those crops on the farm would be as great or greater than now.

The place of corn crops in the rotation, as utilisers of the manurial residues of forage crops, and the value of their straw as food and as litter, make them indispensable, to say nothing of the returns from the use of the grain as food, or the sale of both grain and straw when they could be more profitably turned to account that way.

In spite of the low prices of grain, I am certain that the culture of straw crops will continue to hold its own, and perhaps even increase, unless some great and unexpected change comes over the agriculture of this country. We import enormous quantities of feeding-stuffs for our live-stock, and, as farmers begin to realise how to mix and use their home-grown corn for feeding purposes, they will find that it will pay them better to use increasing quantities of their own corn at £6 a ton than to buy cakes at £8 or more. It is easily possible by good husbandry to produce a ton of grain of any kind to the acre, and if we can realise a minimum of £6 per acre for corn, besides the straw and chaff, there is some prospect of growing corn at a profit. I have already pointed out that we must lay our plans to produce the largest crop possible, with the greatest economy consistent with good farming. The low yield of wheat in most countries is due to bad farming rather than to poor soil.

I have more than once heard the late Sir John Gilbert

speak with withering contempt of the foul and badly-farmed condition of much of the land in the United States, when pointing out that with all their magnificent virgin soils, the average yield of wheat in that country was little more than that produced at Rothamsted, on the land which had grown wheat for over fifty consecutive years without the application of any manure whatever.

The average yield of wheat in this country is between 29 and 30 bushels per acre, but there is no doubt it might be largely increased. We must not suppose, however, that the yield of corn of any kind is capable of indefinite increase, for until some new and entirely different varieties are introduced, a point is soon reached at which the plants so shade each other that proper development cannot take place. It is the experience of practical men, and it has been found on the heavily manured plots at Rothamsted, that in a season of much sunshine and moderate rainfall, yields of from 55 to 65 bushels can be obtained, whereas in a showery sunless season the most promising pieces of wheat fail to stand upright; consequently the ears cannot develop properly and the yield will be poor. In spite of this, wheat remains the most certain crop a farmer can grow, for with good cultivation good crops can practically always be obtained.

Wheat can be grown on almost any soil and after any crop, but being a plant which makes its greatest growth early in the summer, before nitrification has reached its maximum, it especially requires a fair supply of available nitrogen to be present in the soil. It therefore succeeds well after a heavily manured crop, such as potatoes or mangolds, as it has an opportunity of using the nitrates formed during the autumn from the unexhausted manure. It also thrives well after any of the nitrogen-producing crops; and any tillage which encourages nitrification, such as a bare or summer fallow, is an excellent preparation.

In manuring for wheat this matter of the nitrogen-supply must be carefully borne in mind; and in cases where the land is poor, and the preparation not suitable, we must adjust our manures to supply this autumn and spring nitrogen. The autumn supply is best provided by a small dressing of well-rotted dung, or by a hundredweight of sulphate of ammonia per acre, but the spring supply is most safely secured by about a hundredweight of nitrate of soda put on as a top dressing in April, although it should not be always necessary to apply both to the same crop.

Although the extensive root system of wheat enables it to obtain with comparative ease the phosphates and potash it requires, I have found on certain soils deficient in phosphates, particularly on the Boulder Clays, that applications of nitrogenous manures alone produced a large increase in the straw with a very small increase in the corn; but, through the addition of a small dressing of superphosphate in the autumn, say two or three hundredweights per acre, the nitrogen was able to give a considerable increase in yield as well as in straw. On farms on which roots are extensively grown with phosphatic manures, the application of phosphates to wheat should never be necessary.

The most suitable time for sowing wheat is during October, but sowings may be made with success from the middle of September to the end of November, although with the late-sown crops the weather during the winter is the principal determining factor. I have had splendid heavy-yielding crops from September sowings, and the sample which won the first prize for white wheat at the Royal Show at Windsor in 1889 was sown on the 1st of December.

The quantity of wheat required to sow an acre depends much on circumstances. One bushel may be sufficient to drill on a well-tilled fallow in September, although I do not recommend such small seedings. Up to the end of October 2 to $2\frac{1}{2}$ bushels are ample, but for later sowings 3 bushels are required. The rule is, the earlier the sowing and the better the condition of the land, the less the quantity of seed required, and *vice versâ*. When seed is sown early on good land, the plants will always stool out well, so that several stalks are produced from each seed, and there is also less danger of accident to the young plants.

The worst enemy of wheat is the rook. Although wheat is subject to the attacks of a multitude of different kinds of fungus and animal pests, their destruction is as a drop in a bucket compared with those of the rascally rook. He digs up the grain with his powerful beak as soon as it is sown, and continues to do so till it appears above the ground ; then he ceases digging and pulls up the plants by catching hold of the shoot in his beak, because every now and then a plant comes up with the kernel still attached, which he can eat. There is nothing so annoying as to see whole rows every plant of which is pulled up and left lying on the top of the ground. It is useless to plead excuses about insects, for no possible number of insects would equal the damage of half-a-dozen rooks left undisturbed, and the only prevention is constantly to watch with a loaded gun. Then just before harvest he is at it again, picking the green kernels out of the ears. It may be my misfortune, but I never yet found an insect in a rook's stomach, although I remember once shooting a carrion crow and a rook at the same shot, and whereas the rook had nothing but wheat in him, the crow had a number of beetles as well.

Wheat should always be well rolled in the spring, and, if the ground is at all caked on the top, should be harrowed afterwards. As a rule there is little necessity to hoe this crop if the plant is fairly strong and the land not foul, but there are times when thistles and charlock are abundant, and then hoeing might be recommended.

Wheat is best cut early, that is, before it is quite dead ripe, for the quality of the grain is greatly improved thereby. The miller is anxious to obtain wheat containing plenty of gluten, and a wheat cut before it is quite ripe is more glutinous than a wheat allowed thoroughly to ripen; but the farmer has to see that he does not cut so soon as to prevent the grain coming to its greatest weight.

Although it has been proved that other things are necessary for the production of first-class flour, much attention has been directed lately to this matter of the gluten content of wheat, and I should like to point out the object of the miller and baker in desiring such wheat. It seems to me peculiar. It is not because the bread from such a wheat will be nicer, cheaper, or more nourishing, but because a *larger-looking* loaf can be produced from the same weight of flour. The gases given off by the yeast will blow larger bubbles in the dough made from glutinous flour, and so, when we are cutting one of those highly objectionable loaves full of huge cavities, we know that we have got a loaf made

H

from a glutinous wheat, and looking very much larger than it really is. It seems strange that so feeble an object should make three or four shillings difference in the value of a quarter of wheat.

It is difficult to recommend any varieties of wheat as being the best for quality or for yield; so much depends on the district and on the season. It is an excellent practice to try new varieties for a couple of seasons, but it is not my experience that it pays to change the seed every year, either with wheat or any other corn. For high farming the short-strawed varieties are the best.

Besides insects, wheat suffers from several fungoid diseases, but we can prevent two of these; they are smut and bunt. The latter is the most common, and the remedy is to dress the seed before sowing with a solution of blue vitriol (sulphate of copper); $1\frac{1}{2}$ lb. dissolved in 2 gallons of water being sufficient for a quarter of wheat. This is far and away the best of all the various pickles, and this treatment of the seed should never be neglected.

The conditions required for the production of barley of high quality are very different from those required for wheat. The object is different, for whereas in the wheat we want a hard flinty structure with a high nitrogen (gluten) content, which is brought about by rapid ripening, in barley, for malting purposes, we require a soft, white, starchy structure with the minimum of nitrogenous matter, conditions which are brought about by slow maturation and thorough ripening. The ripening of the best wheat is forced: in Northern America and Russia by extreme sunshine and heat, and by the rapid drying up of moisture; in this country by early cutting. The best malting barleys are produced naturally where the climatic conditions are not so extreme, and on light soils, where the food-supply is rapidly made available in spring, and ripening is brought about by the gradual exhaustion of the supply of nitrogen in the soil, as well as by the sun, without extreme loss of moisture. Dry weather during harvest is, of course, essential to production of the best coloured barleys, but, I think, from what I have said, we can trace out the means of assisting the conditions conducive to the production of large crops of barley of fine quality.

The first essential is, then, rapid growth in the spring and early summer, and for this we require thorough cultivation and an abundant food-supply. The land must be clean, and deeply ploughed sufficiently early to allow of its thorough pulverisation, either by frost or by mechanical means. A fine deep tilth is essential, and if the land is a little lumpy, for instance after sheep, it is better to take time to produce a tilth by rolling and harrowing than to sow the barley a fortnight earlier amongst the clots. Barley is a shallow-rooted plant and requires its soil food in a readily available condition in the upper layers of soil. This, together with the fine moist condition of the deep tilth, gives us that rapid growth we are anxious to obtain, for every grower knows that barley once checked in its growth never produces a first-class crop.

Turnips grown with superphosphate and sheeped off give splendid conditions of manuring, owing to the residue of the phosphates and the supply of available nitrogen. Barley should never be directly manured with dung, and if the crop requires manurial assistance it should always be by means of soluble chemical manures.

An abundance of available phosphates in the soil is necessary for the production of fine large starchy grains, and this is best supplied by about 3 cwts. of super-phosphate per acre harrowed in with the seed. Potash is often very useful for barley, and, on soils which do not actually require it, it stiffens the straw and produces a paler-coloured sample. I have known the addition of muriate of potash to a manure raise the price of a sample by 4s. a quarter. Half a hundredweight of muriate or sulphate of potash is sufficient.

Except after a crop eaten on the ground by sheep, nitrogen is nearly always required by barley in its early stages of growth. Nitrogenous manures should never be applied to barley without phosphates, as by themselves they produce a thin and flinty kernel; but with phosphates, 1 cwt. to $1\frac{1}{2}$ cwt. of nitrate of soda on heavy land, sown as soon as the barley peeps through the ground, or on lighter land 1 cwt. of sulphate of ammonia harrowed in with the seed, will often produce great increases in the crop without spoiling the quality. A good general barley manure, such as is often sold at £7 or £8 a ton, can be made by mixing 5 cwts. of superphosphate, 3 cwts. of kainite, and 2 cwts. of sulphate of ammonia, and applying the mixture at the time of seeding at the rate of 4 or 5 cwts. per acre. It is worthy of remark that common salt often produces an increase in a crop taken after mangolds, and it is commonly applied for this purpose in Norfolk. Care must be taken not to manure barley too heavily, especially with nitrogenous manures, and no manures should ever be drilled in with the seed, but always sown broadcast, and harrowed in if possible.

There has been much discussion during the last few years as to the best date at which to sow barley. I have seen crops harvested which have been sown at all times from the last week in December to the first week in June. Winter barleys should of course be sown in September or October, but I am not speaking of these. March is, in my opinion, the proper month to sow barley; the sowing may be begun during good weather in February, and it may be continued throughout April, and excellent results obtained. I am inclined to think the advocates of very early sowing will receive a considerable check the first sharp spring frost we get, for we have not had a severe spring since 1895. I may be unduly pessimistic, but I cannot help thinking that many of the new varieties of winter barley and oats introduced during the last five or six years will succumb to the first hard winter we get, for I have noticed that they are severely thinned even by the frosts of the mild winters we have had lately.

The date of sowing has, I believe, a considerable effect on that pest of all barley-growers, the fungoid disease known as smut.

Fungologists tell us there are two kinds of smut—the open variety, in which the black spores are all exposed, and usually get blown away before harvest; and the closed variety, where the spores are enclosed in a covering retaining the shape of the barley ear, and are consequently carried and thrashed with the barley, often causing a considerable reduction in the value of a sample. I have observed that the earlier sowings of barley are principally affected with the open variety, while late sowings invariably contain the closed variety, and the later the sowing the greater the amount of smut, especially if the germination of the seed in the ground is very rapid. A process of scalding with hot water as a preventive has been recommended by Jensen, for barley cannot be successfully dressed like wheat, but with fairly early sowing I hardly think it is necessary.

Barley should always be drilled, both to ensure even germination and to enable it to be hoed if very weedy.

The same rules apply to barley as to wheat with respect to the quantity of seed required ; the better the conditions, the less the seed which need be sown. A number of experiments in Norfolk showed that, on the average, 2 bushels per acre gave the best results, although there was very little difference between that and $2\frac{1}{2}$ bushels. Many growers of fine barleys sow very thickly to prevent stooling, for they say that the side shoots are later at harvest than the main stem, and so the ripening of the crop is uneven.

Many a good sample of barley is spoilt by mistakes in harvesting and thrashing. Barley should never be cut till it is dead ripe, and it must not be allowed to stand after it is ripe, or it will black on the ends of the grain. In the eastern counties a large acreage is mown with the scythe and carried loose, but if it is clean it may be cut and tied by the binder. All laid and damaged patches should be carried by themselves and never mixed with the best. It should be carried as soon as dry, and should stand at least six weeks in the stack, for all corn sweats a little when first stacked. When thrashing, the drum of the thrasher must not be set too close, or it may crack the kernels, and the hummeller, or horner, which knocks off the awns, must be adjusted so that it does not skin the end of the grain; but this depends on the condition of the grain as to dryness. Malting barleys should be thoroughly winnowed to get rid of all small and badly-developed grains, for they can be used for grinding, and their presence depreciates the value of a sample.

It is almost impossible to recommend varieties, but seed should be chosen which is large, plump, and thin-skinned. Lean barleys with thick, loose skins should be avoided, although a good sample discoloured by the weather may be used. I believe it is possible with care to produce good high-class barleys on land which cannot be considered naturally suitable for them, and prizes have frequently been won by barleys grown on heavy land.

Barley of the Chevalier type is usually the best in quality, if not in yield; but where malting barleys are not desired or cannot be grown, then some of the heavier yielding and coarser kinds may be grown for home consumption. It is a valuable food-stuff for use on the farm.

Malt used to be much used as food for sheep, but, strangely enough, as soon as the great agitation had succeeded in getting the Malt Tax repealed, it went out of fashion, and one seldom now hears of it being so used. It possesses no great advantage over barley.

Oats are much more extensively cultivated in this country than they used to be. Their price has not fallen to the same extent as wheat; the yield is so much larger that the returns per acre may be greater; the grain is a splendid food, and the straw is a valuable fodder. I once grew 11 quarters to the acre, and yields of 12 or even 13 quarters are not unknown. The oat is a vigorous feeder, and seems to abstract from the soil every available atom of food that it can find. No corn crop responds more to manurial treatment than oats, and generous treatment is required for large crops. For its full development the oat requires a moist climate, and for this reason it is much more grown in the north and west than barley. Although deep cultivation and a good tilth are as useful to the oat as to any other crop, it is more independent of them than barley, and with plenty of soil food and rain, it will flourish amongst clots and weeds which would be fatal to the more delicate barley. Oats may follow any other crop, and the usual cultivation is a simple ploughing and harrowing.

On stock farms winter oats may well take the place of some of the wheat, and coming to harvest very early, often early in July, there is a chance of following them immediately with a catch crop such as white turnips, mustard, or rape.

The gray winter oat is fairly hardy, and there are several varieties of black oats which are equally so. These varieties should be sown in the end of September or early in October, and should be treated in much the same way as wheat. It is advisable to test the germination of oats before sowing them, as they often contain many immature grains and are liable to damage in the stack by heating. A ready way of doing this is to place a handful of sand in a saucer, damp it, and sprinkle an ascertained number of grains, say 200, on the top. Cover the grains with a pad of wet blotting-paper, or the saucer with a sheet of glass, and place it in a warm place. In three or four days the little white roots should appear, and as each root appears that grain should be removed till none are left but the infertile. Care must be taken that all the grains are equally damp and that they are not allowed to dry up or the result will be useless. Gray winter oats often contain the seeds of a large grass, *Bromus secalinus*, which, although not apparent on casually looking at a sample, can be easily seen on careful examination. A sheaf of this grass was brought to the Agricultural Department at Cambridge with the assertion that a whole field of winter oats had turned into it during the winter, and the case was given to me to investigate. Fortunately I was able to obtain a sample of the oats sown, and I found it contained 20 per cent. of this grass seed; yet neither the farmer nor the merchant had noticed it. The explanation was simple: the rather severe winter had killed the bulk of the oats, but the grass had survived.

The varieties of spring oats are legion, and their numbers increase year by year. They are divisible into two main groups—the Tartarian, having the spikelets of the head leaning in one direction, and the Common oats, with a spreading head. Of each of these there are black and white varieties, and one newly introduced variety midway between the Tartarian and the common.

Many experiments have been carried out lately in testing the varieties one against another, and although there are most extraordinary differences in the yields, the only possible conclusion seems to be that every farmer must test for his own farm and farming. It is certainly worth his while to do so, for differences of 20 bushels or more may be found under exactly similar conditions.

Tartarian oats take longer to mature than the spreading varieties and should therefore be sown earlier. Spring oats may be sown from February to May, but, as with barley, March is the best month. The quantity of seed required per acre is a much-discussed question, the practice varying from 2 to 8 bushels per acre. It has been repeatedly shown by experiment that these extremes do not yield so well as medium quantities— from 3 to 4 bushels. The large seedings are always found in late and damp districts, where immature, badly developed and damaged grain would be more frequent, so that to the rules given for regulation of the seeding of wheat and barley may be added for oats, the dependence on the percentage of grains which will germinate.

Oats do not easily become laid by rain when not excessively thick, and therefore will stand heavy manuring. Sulphate of ammonia and particularly nitrate of soda may be used more freely than for either wheat or barley, 1½ cwt. of the ammonia or 2 cwts. of the nitrate, if required; but care must be taken that an abundance of phosphate is also available, or the straw will be increased without a corresponding increase in the corn. In some districts a little kainite is useful with oats.

Although not injured by fungoid diseases to any extent, oats are often affected by insects, particularly by the wire-worm. This creature, the larva of the click-beetle, is the cause of great loss to farmers, especially on the ligher soils. It attacks every kind of farm crop, except, perhaps, mustard, beginning on the seed as soon as it is deposited in the ground, and feeding on the underground stem or roots until the crop is harvested. There is no possible method of destroying the insect in the ground which will not also destroy the crop, and palliative measures only can be adopted. I state this confidently, for up to the present none of the cures one hears of have ever been found to work when tested by practical experiment.

Rolling with a heavy ribbed roller across the rows may have some effect in making migration less easy, and I always practise this myself, but I think its effect is as much due to its help to the plant as its discouragement of the wire-worm. The most effectual treatment, however, in my opinion, is to help the plants by thorough cultivation and judicious manuring. It takes the wire-worm longer to finish the consumption of a large healthy plant than a small and weakly one, and while he is engaged on one (for he does not shift about much) the other plants have a chance to become larger and less likely to succumb to his attacks. It is when the growth of a crop seems to be stagnant that the wire-worm does most injury, and a dressing of nitrate of soda, owing to its rapid action, is most likely to start the plants into rapid growth.

Oats should always be cut early, very early, in fact as soon as they begin to look yellow. They must of course stand longer in the shock or stook, but they will mature gradually, and a brighter, thinner-skinned grain will result, while the straw will be infinitely better as fodder.

Rye is now very little grown in this country as a cereal crop, although it is extensively used in the light-land districts as a catch crop for sheep feeding, and the acreage still harvested is largely used for providing seed for that purpose. The straw of rye was once eagerly bought by harness-makers for stuffing collars, but now there seems to be little demand for it.

For harvesting, about 2 bushels per acre are drilled in September or October, and the crop is treated in the same manner as wheat. It ripens early, but the yield is small—3 or 4 quarters per acre—and the price is low.

Its immunity from the effects of frost and its exceedingly rapid growth in the spring, however, make it a valuable crop for providing early feed for sheep. On light-land sheep farms it is sown in August or September and is fit to turn the sheep on to in April or early in May. The stubble of the previous crop may receive a shallow ploughing with a double-furrow plough, and the rye sown broadcast or drilled at the rate of 2½ or 3 bushels per acre. After sheeping off the rye the land may be prepared for turnips. Rye may be sown at almost any time of year for feed, but it is in the spring that it is more useful than almost any other crop. From a half to one bushel per acre is often sown mixed with tares for feed purposes, but I prefer winter oats to rye for this. Market gardeners often use rows of rye three or four feet apart to protect their more delicate plants from frost and wind.

The crops with which we now have to deal are those leguminous plants which are grown for their seed— beans and peas, and, to a less extent, tares or vetches. The value of these, together with the forage plants of the clover class, can hardly be overestimated as assistants to profitable agriculture. Their frequent occurrence in the rotation of the farm crops not only gives us the means of enriching our soil in nitrogen, but provides for the use of our stock large supplies of nitrogenous "flesh-forming" matter which is so deficient in the food supplied by the cereals.

The value of these plants in enriching the soil was recognised by the Romans, and in spite of the efforts of the more modern agriculturist to prove its impossibility, it was believed in by farmers generally, up to the time when Hellriegel was able to tell us why, and by what, the soil was enriched.

Although we now know that, even if the whole crop be removed, the soil is still richer in nitrogen than it was before, we must remember that it is only in nitrogen that the soil is enriched, and that a very much larger proportion of mineral matter (phosphates, potash, lime, &c.) is removed than in the case of cereal crops. In soils at all deficient in these ingredients this abstraction must be remembered and provided for.

I must also point out a very serious error that has crept into many recent agricultural writings on this subject. Because these plants are able to obtain their nitrogen from the air through the bacteria, and can grow and produce a large yield under conditions in which they must obtain all their nitrogen from the air, it has been inferred that nitrogenous manures will not benefit the crop. This is a very great mistake, for although it would usually be a waste to apply such manures to these crops, there are times when a little available nitrogen may be applied with advantage; and my own experience and the experiments of others have shown that the crop can often be greatly increased thereby.

Of the leguminous crops grown for corn, beans are undoubtedly the most important. The extension of the area given to their cultivation is unfortunately checked by the fact that they are very uncertain in their yield. They are very subject to the attacks of birds, both when first sown and when appearing through the ground; they must be kept clean by hoeing; they suffer from attacks of the bean aphis, and are very dependent on the influences of the season. In spite of these drawbacks, beans deserve to be cultivated on a larger area than the 242,000 acres devoted to them last year in the United Kingdom.

There are several varieties of beans, but the same amount of attention does not appear to have been given to the introduction of new varieties for field cultivation as has been given to most farm crops; indeed it is not easy to obtain seed pure, and by name. Commonly one only hears of winter and spring beans; but the latter may be divided into three classes—the common spring or horse bean; the mazagan, which is much larger; and the tick bean, which is small and very round. I have grown all four kinds, but I find it impossible to state which is the best, except that the winter bean is much the most reliable in yield, and for that reason should be the most extensively grown.

Beans must be sown early if a good yield is to be expected, the winter beans in October and the spring varieties in February. In planting, beans are often ploughed in, being sown in every alternate furrow by a little drill fixed on the plough, the furrow-slice being turned over on to the top of them. With a not too solid furrow-slice about three inches deep, the beans come up well, and are protected from the ravages of the rook, but, where it can be managed, drilling two or three inches deep is undoubtedly the best way. The quantity of seed required varies from 2 to 4 bushels according to the size and quality of the seed and the time and condition of the soil.

In old times dibbing was the common method of planting beans, and a very excellent one too, but unfortunately it is now difficult to obtain men who can dib. In the days when children were available—fortunately over now—the man used to make the holes with a dib in each hand, and two small children followed and dropped the beans into the holes. Five beans for each hole was considered the proper number—

> " One for the pigeon, one for the crow,
> One to die, and two to grow."

The holes, being six inches apart in the row, gave room for the hoe to be swept between the groups and facilitated the cleaning from weeds, and, although the cost per acre is more than drilling, less seed is required, and I have experienced many satisfactory results from dibbing.

The proper distance between the rows of beans is a much-disputed question. I have seen beans drilled with the rows from 8 inches to 28 inches apart ; the narrow rows being adopted as an attempt to do without hoeing, and the widest being used to give greater facility in horse-hoeing. Both are, however, extremes, my own experience and observation pointing to rows from 16 to 20 inches apart. Beans must be hoed, but a good deal of labour may be saved in this way. In the spring, when they are about 2 inches high, and the seedling weeds are beginning to make their appearance, they may be well harrowed with a " Parmiter " chain harrow or an American weeder on a dry day, when millions of small weeds will be destroyed without damaging the beans at all. In ten days or so the horse-hoe can commence operations, a second hoeing to be given just before the beans get too big to allow the passage of the horse. Hand-hoeing can often be dispensed with altogether, or may be confined to cleaning among the beans in the row; and where the rows are not too far apart, the bushy tops of the beans will soon prevent the growth of weeds. Where beans are regarded as a cleaning crop, the Scotch method of growing them on raised drills or ridges, in exactly the same way as turnips, is a good one, as the horse operations between the rows is greatly facilitated.

Nothing seems to do beans so much good, as a rule, as a good dressing of partially rotted farmyard dung ploughed in shortly before the seed is sown. In a number of experiments in Essex it has been found that the addition of 3 or 4 cwts. of superphosphate to the dung has considerably increased the yield, and, in cases where lime is deficient in the soil, a dressing of that substance has been found to double the crop. Strangely enough, although the bean takes large quantities of potash from the soil, dressings of potash manures have not been found to produce a profitable result. The bean prefers a stiff, compact soil, growing well on all clays, but luxuriating in a well-farmed clay loam. Beans sown early on well-tilled land may be expected to yield from 4 to 6 quarters per acre, but the winter bean is the more prolific, and less subject to injury than the spring varieties. It is seldom the winter bean is injured to any serious extent by the black aphis, as it has practically finished its growth before this appears; but I have seen a promising field of spring beans reduced to black stalks by the terrible pest in less than three weeks. Its first appearance is indicated by a few stalks becoming blackened by masses of the aphis, and from these centres it spreads with marvellous rapidity. I believe that the attack can often be mitigated if taken in time, and these isolated stalks gathered into pails, to be carried off the field and burnt.

There is much difference of opinion and practice as to the correct time to harvest beans. In many parts of the country they are cut as soon as the leaves begin to turn black, and while the stalk and pods are still quite green ; in others the crop is allowed to remain until the whole plant is black. I have cut them both ways myself, and am unable to state that I have observed any difference in the sample; but, as beans should always be cut with the binder whenever possible, there is less fear of their being knocked out if they are cut while the pods are still a little green. Beans should be put up in small shocks to get well dry before carrying, but I have noticed that, if thoroughly sun-dried, a few greenish pods will make no difference in the stack, unless they are required for immediate thrashing.

Peas are more commonly grown on the lighter soils not suitable for beans, but they will grow equally well upon heavy land. They are unfortunately the most uncertain crop that a farmer can grow, for although all crops are somewhat of a speculation, peas are more risky than South African mines.

I remember seeing a stack of peas thrashed, which yielded nearly 7 quarters to the acre, and they were sold at 50s. a quarter. The next year that farmer sowed 70 acres of these same peas, and the yield was 1½ quarters per acre. It must have been the knowledge of this uncertainty that caused an old farmer to reply, when I

I

remarked on his large acreage of peas, "Yes, and there are only two kinds of people who grow peas—gentlemen and fools; and I know I'm not a gentleman."

In spite of this uncertainty, peas are often a very useful crop on light land, and there is no doubt that on well-farmed land the uncertainty is reduced considerably. There are many varieties of peas which can be grown on the farm, gray or dun, maple (speckled), white, blue, and green. For corn purposes all but the green are commonly grown, and perhaps the maple is the best yielder, but much depends on the soil and climate.

The white, blue, and green varieties are often grown near large towns for picking green, and there is no doubt that the return from such a crop is fairly good; the haulms of the peas, being cut as soon as the pods are gathered, make excellent hay.

For field purposes peas are usually drilled at the rate of from 2 to 3 bushels per acre in February or March, in rows from 16 to 24 inches apart, according to the variety, some producing a more vigorous growth of haulm than others. They are kept clean as recommended for beans, but the horse-hoeing must be stopped early owing to the peas intertwining across the rows. Unfortunately peas never cover the ground closely enough to prevent the growth of weeds, so that, however clean they may be at the last hoeing, there are sure to be weeds at harvest. Peas cannot be cut by any machine, so the scythe or hook is usually employed, and as they are cut they are made up into little cocks or wads. As peas are always ripe before the other corn crops, the slowness of this operation does not interfere with the harvesting of the other crops.

It is a very common, and a fairly good practice on the

heavier soils, to sow spring beans and peas together, say 2 bushels of beans and 1 bushel of peas. The beans hold the peas up, so that their individual yield is increased, while the beans do not seem to suffer, and the crop is made more certain and is more easily harvested.

Peas seem to do best, on land in fairly good condition, without the direct application of manure. They may receive a dressing of rotted dung, and it has been found by experiment with the garden varieties for picking green that a dressing of superphosphate makes the yield larger and earlier.

Tares or vetches are principally grown as a forage crop, although a considerable acreage is harvested for seed purposes. The seed has a high feeding value, and often commands very remunerative prices; but the yield is not great, running from 3 to 4 quarters per acre. No agricultural product varies more in price than tares; I have known them from 28s. to 64s. a quarter. There are several varieties, but only two are generally recognised, the winter and the spring—the seed of the latter mostly being imported from abroad.

Winter tares should be sown in October, the land being simply ploughed, and the tares drilled at the rate of about 3 bushels per acre. Pigeons will come for miles to a field newly sown with tares, but if watched for a time the crop requires no further attention till ready for use. So densely do tares cover the ground that it is very rarely any annual weeds are ever seen amongst them. When harvested for the seed they are allowed to remain until the bottom pods are ripe, and although they are green at the top they are cut and dealt with in the same way as peas.

As a forage crop there is nothing to beat the tare: it

is highly nutritious, produces a large weight to the acre, every kind of farm animal is fond of it, and it is a good preparation for any other crop.

For hay the tares should be cut when they are in full flower, and just as the lower pods are beginning to show. From 2 to 3 tons an acre can be obtained, but great care has to be taken in getting it perfectly dry, or it is apt to mould. This hay is excellent for horses, sheep, and milch cows, as it contains a large percentage of nitrogenous substance. For soiling purposes—that is, for carting green to the animals—the cutting should be begun early, as tares are indigestible when the pods become hard. Horses, cattle, sheep, and pigs will thrive on a diet of green tares; but in carting, care must be taken not to place the fodder in large heaps, or fermentation will take place in a few hours, and the quality of the food be spoilt.

Tares are also largely consumed on the land by sheep, as they provide early and nutritious feed besides heavily manuring the soil.

For sheeping off it is usual to sow with the tares about a bushel of rye or winter oats, which keep the tares up and allow the feeding to be begun earlier. Where early feed is of importance to the flockmaster, it is usual to manure at any rate some of the land on which the tares mixed with rye or oats are sown; and considering the manuring the land will get by the sheep, I think that 3 cwts. of superphosphate put on with the seed, and 1 to $1\frac{1}{2}$ cwt. of nitrate of soda in March, will be likely to produce a quicker growth than dung, and be better for the farm. Always begin sheeping early, for the crop should be finished by the time it is in full flower. Where rye has been used

THE EVENING

with the tares, it soon becomes too hard for the sheep to eat, and for this reason oats are preferred.

As all winter tares not wanted for seed should be removed by the end of June, there is plenty of time to give the land a good cleaning if required, or to plough it and obtain a crop of turnips. Spring tares can be sown at any time from March to the beginning of June, and will provide a succession of feed right through the summer.

I have found tares and oats make capital ensilage, and I am convinced that there is no forage crop we have which will provide such a succession of food, or is so valuable where a large quantity of stock is kept.

As part of my idea of making the farm as self-supporting as possible, I should like to draw attention to the cultivation of linseed, the produce of the flax plant, which has not, I think, received the attention it deserves, although I have found it cultivated in small quantities for its seed.

The grain of our cereal crops, wheat, barley, and oats, supply us with abundance of starchy matter for the use of our farm stock; beans, peas, and tares give us large quantities of nitrogenous flesh-forming substances; but in oil, a necessary ingredient of all foods, our usual farm crops are very deficient, oats being the richest with only 6 per cent., while beans contain only $1\frac{1}{2}$ and wheat 2 per cent. Now linseed contains from 33 to 34 per cent. of oil, so that a very small proportion of this added to our other foods will easily bring them up to the most nutritious proportions.

Flax grows splendidly in Great Britain and on almost any class of soil, provided it is free from weeds and in fair condition. Its cultivation is very simple. The land is ploughed early so that a fine tilth is produced, and in the end of March or beginning of April a stoke of the harrows is given and the seed drilled at the rate of not more than half a bushel per acre. The seed should not be deposited more than an inch in depth, and the rows should be about 12 inches apart. A light rolling once over with the lightest possible harrow should be sufficient, for it is important not to bury the seed too deeply. One hoeing either by horse or by hand should complete the culture.

The flowering of the flax takes place over a considerable period, so that its ripening is also irregular, but it is generally found best to begin harvesting as soon as the first-formed capsules are fully ripe. In districts near linen or flax-spinning factories the straw is exceedingly valuable, and it is usual where it can be sold to pull the plants up by hand and tie them in small bundles. Unfortunately, high railway charges prevent the straw being sent far, so that in many cases the value of the straw is disregarded. Here the crop may be cut with the scythe or reaping machine and tied up into sheaves. I have never used or seen the binder used for this crop, for the acreages within my experience have always been small, but I see no reason why this implement should not be used on dewy mornings or on a dull day, care being taken to tie the sheaves as small as possible. After standing in shocks till dry, the crop is carried, stacked, and thrashed in the usual manner. The yield is from $2\frac{1}{2}$ to 4 quarters per acre, and although the straw has no feeding value, the chaff is excellent food. The small, dark-coloured seed is the better to use, being superior to the larger, light-coloured

Indian seed. The chief enemy of the flax is the small bird, and the finches of various kinds are apt to do a good deal of harm at harvest if not watched. As linseed varies from 45s. to 55s. a quarter, it is sometimes a valuable crop though small.

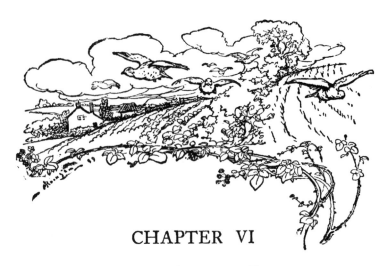

CHAPTER VI

Roots and Green Crops

WE have now to consider those crops which are grown almost entirely for consumption on the farm, and the cultivation of which allows a considerable amount of cleaning or fallowing to be done in the intervals between their growth and that of the preceding crop. There are many which come under this heading, far more than are commonly grown, but the root crops, mangolds, swedes, and turnips, take the first place.

Although these crops are of great importance in the economy of the farm, yet I must protest against the excessive and often wasteful area devoted to their culture in many parts of the country, where a fourth of the cultivated area of the farm is often occupied by them. It is impossible to breed or keep on the farm sufficient stock to consume such an acreage of roots, even if the crop be small, and large numbers of sheep and beasts have to be bought on purpose to consume these roots, even though it may be known that there is no chance of selling them again at a price which will pay for the labour and cake, let alone the cost of the roots. Surely it is a waste to grow such an extent of so expensive a crop merely to manure the land, for if the crops following can pay for such expensive manuring, then the profit on corn-growing must indeed be great. It is my contention, and many agree with me, that if this area were devoted to provide crops which could be used or fed off all the year round instead of only in winter, a larger flock might be kept, the fallowing operations spread over a longer period, and the manurial result largely improved.

Until a comparatively few years ago the turnip was considered a resting crop, and was even supposed to obtain much of its nitrogen from the air; but this has been conclusively proved to be fallacious. Not only does the turnip draw no nitrogen from the air, but a crop of even 17 tons of turnips will remove from the soil $2\frac{1}{2}$ times as much nitrogen, 5 times as much potash, and $1\frac{1}{2}$ times as much phosphoric acid as a crop of 30 bushels of wheat with its straw. Would it not be much better to grow larger areas of crops which will *enrich* our soil in nitrogen? A considerable quantity of roots have become almost essential for stock-keeping, but surely we can grow the requisite quantity of roots on a smaller area.

The Agricultural Returns give the average yield of turnips and swedes in Great Britain, for the ten years ending 1901, at $12\frac{3}{4}$ tons per acre. Now an acre of turnips, planted in rows 27 inches apart and with 9 inches between the plants in the row, and in which every turnip weighed only 1 lb., would yield $11\frac{1}{2}$ tons.

Can we not then grow better crops than an average of 12¾ tons per acre? As a matter of fact, twice or three times this crop can easily be grown where sufficient care is taken and the labour and attention not spread over too large an area; it is a regular plant without gaps that is required for a large yield, rather than enormous individual roots.

That we have plenty of time in most cases to clean our land before we need sow the roots is of course an important factor, but it must be very clearly borne in mind that there are often times when we cannot obtain both a clean fallow and a good crop of roots. This is very commonly the case in loams and the stiffer soils, and many a farmer "runs after two hares and catches neither." If the land requires cleaning, clean it, and let the roots take their chance; if roots are a necessity, then select a piece of fairly clean land and grow roots on it.

The cultivation for roots should always begin with a thorough tearing up of the stubble with the cultivator in the autumn, and all the weeds which can be got out should be gathered up and burnt. The weed seeds left from the preceding crop will now begin to germinate, and can be destroyed by subsequent tillage, but what that tillage should be depends entirely on the nature of the land.

Some soils, when ploughed early and reduced to a fine tilth either by the frost or by harrowing, quickly become so sad and sticky underneath that a second ploughing in the spring is not only difficult, but leaves on the surface large putty-like lumps which dry into the consistency of bricks, and a really good tilth can never again be produced. These are the heavy soils, where, if roots are really required, spring cleaning operations should never be attempted. It is surely very evident that the tillage of such land must be very different from that of the light soils, which never become sad, and where the more they are ploughed and harrowed the finer and deeper the tilth becomes. The soils, on which cleaning operations before roots can be undertaken with confidence, are those which are commonly called turnip soils.

The extent of the cleaning and the course to be adopted must of course depend on the soil; but with these remarks we will proceed to consider the root crops separately, taking it for granted that the roots are being grown because they are required, and that as large a crop as possible is desired.

The mangold-wurzel is a most important root, and deserves far wider cultivation than it receives. Not only does it yield a heavier weight per acre than any other root, but it is more nutritious, is relished by all kinds of stock, it is easy to grow, and its keeping powers are remarkable. Mangolds are frequently used for cows and sheep through May and June, and I have used them for my own cows till September.

This root can, however, never replace the swede altogether, for it ripens slowly and is seldom suitable as food for stock before January, while the swede may be used in November if required. The mangold is above all others the root for heavy lands, but it can be grown with equal success on the lightest of soils, though the preparation must be different. On heavy soils the land must be deeply ploughed, after the cultivating already mentioned, not later than January, and any dung to be applied should be ploughed in then. Personally I prefer to use my yard manure for the corn crops and to grow roots with artificial manures alone, but a dressing of twelve or fifteen loads may be given, if it can be spared.

Now comes the secret of success on these soils: do not touch the land again till you are ready to sow the seed. About the third or fourth week in April, choose a dry day, sow the artificial manure broadcast over the surface, harrow in with a fairly heavy implement, and drill in the seed. Once over with the roll, which must never be omitted, finishes the operation.

It will be seen that this precludes any cleaning in the spring, but, if mangolds are wanted, that is the way to get them. One very successful root-grower I know, recommends the steam cultivator instead of the plough for the winter preparation, and it is no doubt an excellent plan where available, but he agrees with me in insisting that the land should not be touched again till it is ready for sowing. On loamy land the spring-tined cultivator may be used in the spring, but never the plough.

On some of the water-logged clays in Essex the land is thrown up in ridges or drills in the autumn, the manure carted on in the winter, and the ridges split back sufficiently early for the frost to renew the surface tilth. Ridges are nearly always used for roots of all kinds on the lightest soils, for they allow the manure to be covered in after the cleaning processes are over, and are certainly much easier to clean with the horse-hoe; but I do not think they have any advantage as far as the crop of roots is concerned. I think, rather, that the hollow made by the dung, especially

when it is a little dry, and the rapid drying of the sides,
militate against the proper growth of the crop.

The idea that the manure, whether farmyard or
artificial, should be immediately underneath the plant
is not only erroneous but mischievous, and many a
plant of roots is destroyed in dry weather by the
manure being drilled in with the seed. The principle
is entirely wrong, for the nature of any plant is to send
its rootlets ramifying through the soil in all directions,
in the case of a mangold or swede for 2 feet or
more round the plant, and yet the manure is often
deposited in a narrow groove immediately underneath it. Rain may wash the manure down, but
never out between the rows where most of the rootlets are. For this same reason the practice of applying nitrate of soda just round the plant is strongly to
be condemned.

On light land a dressing of rotted farmyard manure
is more essential to mangolds than on heavier soils, but
it should not be applied till early spring, and strawy
manure should never be used. The manure may be
ploughed in after the cleaning, or spread between the
ridges, if these are adopted; and in this case the artificial manures should be sown over the dung, so that
they may also be covered in when the ridges are closed.

Dung by itself is not sufficient to produce economically large crops of mangolds, and should always
be supplemented with chemical fertilisers. Nitrogen
seems to be their principal requirement, and it can be
applied profitably in all cases. Potash and common salt
usually produce increases, particularly on light land; and
phosphate usually increases the yield on heavy land,
although on many soils it may not be required with

dung. The influence of phosphates in ripening the crop makes the use of small quantities of it distinctly valuable. With a moderate dressing of dung I should recommend as a general dressing 2 cwts. of superphosphate and 2 cwts. of kainite per acre to be harrowed in at sowing, and 2 cwts. of nitrate of soda to be applied in two top dressings; the first immediately after singling and the second a month later, care being taken to distribute it evenly over the whole surface.

When no farmyard manure is used, the quantities of superphosphate and kainite may be doubled and the top-dressing remain the same, but 1 cwt. of sulphate of ammonia per acre should be applied at seeding with the other manures to replace the nitrogen of the dung. Such a manuring as this is capable of producing on average land a crop of from 30 to 50 tons per acre.

The size of a crop of roots of any kind, produced on land properly manured, seems to be less influenced by largeness of the quantity of manure than by the mechanical condition of the soil, the amount of attention given to the crop at the critical periods of its growth, and the weather.

Persistently dry weather is the greatest enemy to large crops of roots; but it is surprising how its effects can be mitigated by equally persistent hoeing and tillage between the rows. Never allow the surface of the ground to become hard or caked; whether there are any weeds or not, the horse-hoe and the hand-hoe should maintain a coat of fine soil over the surface— an earth mulch, as the Americans call it. The distance between the rows of roots has to be regulated as much by the requirements of the horse-hoe as the requirements of the roots, and the best width seems to be from 24 to 28 inches. It is advisable to mix a little swede or kohl-rabi seed with that of the mangolds when you are drilling, as these plants appear more rapidly and mark out the rows, so that the men can see where to hoe much sooner than with mangolds alone. Six pounds of mangold seed and 4 ounces of the swede or rabi seed per acre should produce a good plant, and the hoeing must commence as soon as the rows can be distinctly seen.

As soon as the plants are large enough to handle (and the sooner the better) they should be singled out. The distance we decide to leave between the plants in the row and the care with which the singling is carried out very largely determine the weight of the crop produced. To obtain a heavy crop, the first requirement is that the plant shall be regular and without gaps; the second, that the plants shall not be too far apart.

If the rows are 2 feet apart, and the plants each 1 foot apart, then we should have 21,780 plants on an acre, and at 4 lb. per root the crop should be nearly 39 tons. If, however, the roots are singled at 18 inches apart, then we should only have 14,520 plants per acre, which, at 4 lb. each, would weigh practically 26 tons. Now at the greater distance apart the roots would un-doubtedly be individually heavier than at the smaller distance; but would they each weigh 2 lb. more, which would be necessary to produce 39 tons? I am certain they would not; and with the small-topped mangolds of the present day I am convinced that from 12 to 14 inches apart is ample for the production of the heaviest crops. Again, a given weight of 5 lb. or 6 lb. roots are sounder, will keep better, and are much handier to use than the same weight of 12 lb. or 16 lb. roots, although

I see no reason why roots weighing 8 lb. or more should not be successfully grown at the distances I have recommended.

Mangolds should be taken up not later than the beginning of November, but much depends on the state of maturity of the crop. As soon as the leaves begin to turn yellow and drop off the crop is ready for lifting; but if the leaves are still a vivid green the crop should be left as long as possible, as it is still immature and growing. A slight frost will not injure mangolds, provided they are not touched while they are frozen. Raising mangolds on frosty mornings often results in much rotting in the pit, and there is a great tendency to ascribe these bad results to causes which had nothing to do with them. The use of artificial manures, particularly nitrate of soda, is often said to prevent mangolds keeping; but many farmers besides myself, who use these manures habitually, know that they have no such injurious results. Mangolds keep the better in long heaps out-of-doors than in a root-house, and I believe that well thatched with dry straw about a foot thick, they suffer less from frost than when covered with earth.

There is a very large number of varieties of the mangold, and it is difficult to say that any one variety is the best, for the kind of soil makes some difference in the comparative value. There is, again, an extraordinary difference in the percentage of sugar in different stocks of the same class. The long red variety is a large growing kind suitable for deep soils, and, although often considered coarse, it usually contains a higher percentage of sugar than any other variety. The globe mangolds grow to a large size and are a good general purpose root, but the different strains vary greatly as to quality. The Tankards are usually of fine quality, and, although they never grow to the great size of some varieties, they should be grown closer together, and will thus produce heavy crops. One of the best varieties for heavy land is shaped like a tankard bulged at the bottom. It is a large cropper, of fair quality, having a small top, and the whole of the bulb growing above ground, it pulls up clean in the most sticky soils.

The action now being taken by the European powers with regard to the Sugar Bounties makes it very probable that the cultivation of sugar-beet may become, before long, an industry in this country.

It has been abundantly proved that we can grow larger crops of beet, containing as high a percentage of sugar as can be produced in any country in Europe. The cultivation is exactly the same as for mangolds, with these exceptions, the plants may be allowed to remain a little closer together in the rows than mangolds, and the last top-dressing of nitrate of soda must be omitted. The tendency of all nitrogenous manures is to prolong the period of growth, and it is essential for a high percentage of sugar that the beet should have finished their growth and have become matured some time before they are taken up. On this account also, phosphates must never be omitted. With a very little encouragement there is no doubt that the cultivation of sugar-beet might become the saving of our heavy lands.

Twenty tons per acre of first-class quality can easily be grown by the method I have stated. The beet would probably be bought at prices regulated by the percentage of sugar contained in samples of the bulk, and might be

K

expected to vary from 15s. to 25s. a ton. To save railway carriage, small crushing and extracting factories would require to be established in suitable districts, so that farmers could cart in their produce and bring back the pulp, which is of considerable value for cattle-feeding. The variation in the price would make it essential that only well-matured roots were grown; but to imagine, as I have seen stated, that a large crop must necessarily be of poor quality, is a great mistake. It is merely a question of variety and a properly balanced food-supply. The phosphates and potash must be in excess of the possible requirements of the plant, but the nitrogen must be so regulated that after causing the rapid growth of the plant in its earlier stages the supply shall gradually become exhausted, causing growth to cease and maturation to take place. These remarks apply to all plants whether in the garden or in the field, and by no nitrogenous manure can the supply be so easily regulated as by nitrate of soda, for the quantity and time of its application control the supply.

Swedes are perhaps the most extensively grown of all our root crops, and they undoubtedly deserve their position. They are the mainstay of the sheep farmer, providing sheep feed from November to April, and they prefer those soils which are most suitable for sheep farming. As with all root crops, the chief requisite for swede-growing is a clean well-tilled soil in an extremely fine tilth. The swede being sown at a much later date than the mangold gives an opportunity for a thorough spring cleaning of the land, which it is too risky to attempt with the mangold. In Scotland and the northern districts of England swedes are drilled fairly early in May, but in the south from the first to the third week in June is preferable, as swedes sown earlier are subject to attacks of mildew.

When dung is used swedes are always grown on ridges or drills, so that the manure may be adequately buried, but it is more common to use artificial manures alone for swedes, especially when the crop is intended for consumption on the land. Although dung alone will produce fair crops of swedes on some land, it is quite inadequate on others, owing to its deficiency in phosphates ; but in any case it is not an economical dressing.

It is fairly well known in practice, and has been abundantly demonstrated by experiments, that a comparatively small dressing of dung, such as 8 or 10 tons per acre, supplemented with from 3 cwts. to 5 cwts. of superphosphate, is more economical and effective than larger dressings.

There is no necessity for dung at all in growing swedes if use can be made of it elsewhere, for just as good crops can be grown without it; a dressing of 5 cwts. of superphosphate and from a half to three-quarters of a hundredweight of sulphate of ammonia per acre will provide all that is required. It is only on land very deficient in it that potash pays on swedes. It is a good practice to grow the swedes intended to be eaten by sheep on the land with artificial manures alone, and to grow those required for carting to the homestead with a small dressing of dung in addition. In this way the land is not unduly robbed, and the following corn crop will be more level.

The dung used for swedes, as indeed for any other crop, should always be partially rotted by having been thrown into a heap for a month or so, and dung straight from the houses never used. Many an unaccountably small crop is due to the use of fresh strawy manure, for it acts by keeping the land hollow, and therefore dry, as well as by causing actual loss of nitrates. Too large a dressing of nitrogenous manure causes swedes to grow coarse and rooty, and to develop a long thick neck between the bulb and the top, as may be observed where swedes grow among mangolds.

Much difficulty is often experienced in growing swedes on heavy land, but it will be found that if the same rules are observed as have been recommended for mangolds, good crops can be obtained. The difficulty lies principally in the matter of obtaining the requisite tilth, and sometimes in the deficiency of heavy lands in phosphates. From 2 to 4 lb. of seed per acre is ample, and, as with all small seeds, care should be taken to buy only the best seed from reliable firms.

In the hoeing, singling, and distance between the rows, the same remarks apply as were made about mangolds, although the distances between the plants is sometimes less.

The greatest enemy of the turnip-grower is the little turnip beetle commonly known as "the fly." Unfortunately there seems to be no sure cure or preventive. There is no doubt, if we could coat the young leaves of the swede or turnip with a substance which is obnoxious to the insect, we might prevent his attacks to some extent, and this is sometimes tried. Fine powders, such as soot, sulphur, or lime, are sown over the plants on dewy mornings, so that the powder may stick to the leaves; and sprays of paraffin have been used with some success. The best thing, however, seems to be to provide the plants with plenty of soluble food, and to drill them under such conditions of moisture and tilth that they will grow with great rapidity. Even with every precaution the whole plant may be cleared off, while at another time not a fly can be seen.

Where swedes are required for cattle in the yards they are usually raised about the end of November, the tops and tails chopped off, and are then carted to the homestead. They keep very much better, like mangolds, in well-covered heaps outside than in a root-house. When required for sheep, however, it is very common to allow them to remain in the ground until consumed. This is a most extravagant, wasteful, and injurious custom, and cannot be too strongly condemned. I have talked the matter over with several flockmasters, and, although all agreed that the practice was bad, their excuse for continuing it was that it saved labour.

All swedes for sheep should be got up in November or the beginning of December, and should be thrown in heaps and covered up with earth in the field. This costs from 8s. to 10s. an acre, but the great advantages are that the swedes all keep sound, become matured and more nutritious, and will keep well without deterioration till the end of April; moreover the sheep never have to eat frozen roots, which is a fruitful cause of abortion and other troubles. It is a common practice to allow a field of swedes to send up their flowering stems in the spring on purpose to provide green food for the lambs. This is a ridiculous practice, for not only are swede tops far inferior in food value to such a crop as thousand-heads, but the bulbs become woody and useless, so that the farmer has been to the trouble and expense of growing swedes only to provide him with feed which he could easily have procured at a quarter of the expense. This all arises from the absurd practice of growing a larger acreage of roots than can possibly be used with economy.

Turnips possess no advantage over swedes, except that they can be sown later and can be used slightly earlier. They are much inferior in feeding value to any other root, and do not keep well. They have a value for sowing on land which could not be got ready in time for swedes, for they may be sown any time in June or July; but that they are ready to use a little earlier than swedes is, in my opinion, counterbalanced by the fact that we can grow many other less expensive crops for feed in the late autumn.

These white and yellow fleshed turnips are of great value on a farm as catch crops—that is, for growing after some crop which has only occupied the ground for a short period; but to grow them over large areas where the more useful swede might have been raised seems to me to be poor practice, although of course there may be conditions which justify their growth.

They require exactly the same conditions as swedes, and may be successfully grown with the manures recommended for that crop. When sown late in the season a fine seed bed and plenty of available plant food are essential, as it is only in this way that rapid growth can be obtained. There are many varieties, some of which are of more rapid growth than others, but the most suitable kind depends largely upon the district. Turnips may be drawn to the stockyards for early use, but are usually consumed on the land by sheep. They are not, however, the best food for sheep, being too watery, and it has frequently been observed that in a good turnip year the fatalities in the lambing-yard are largely increased. The climatic conditions of the northern counties are very favourable to the growth of large crops of turnips, but I cannot recommend them except as catch crops, where better roots can be grown. Swedes, turnips, and other plants belonging to the same order are often subject on light lands to a disease known as "finger-and-toe" or "club-root," which causes the roots to become misshapen in growth, and to rot on reaching maturity. The disease is due to a "slime fungus" which enters the rootlets, and by living and growing there causes abnormal growths and finally decay. This disease is never found on land containing plenty of lime, and there seems to be sufficient evidence to show that dressings of slaked lime at the rate of about 2 tons per acre, applied some time before sowing the seed, will act as a preventive. The galls produced on roots by the grubs of the turnip-gall weevil are sometimes mistaken for true club-root, but there can be no mistake when the little white grubs are found in the excrescences. The damage caused by this weevil seems a negligible quantity.

I now have to speak of three crops, which, although of the greatest possible use in providing green food for our live-stock all the year round, are not cultivated as much as they might be.

These crops are kohl-rabi, cabbages, and thousand-headed kale. Being modifications of the same plant, they require almost exactly the same cultivation and treatment, but their forms and uses are very different.

The kohl-rabi is a most peculiar-looking plant, with its stem swollen into a bulb, on which grow the leaves. Sown at the same time as mangolds, these bulbs mature in September or October, and may either be cut off and stored exactly in the same way as mangolds, or they may be allowed to stand, and be fed off by sheep on the land. Thirty tons or more per acre can be easily grown; they stand the frost well, and as the bulb rests on a short stalk, it can be entirely consumed by the sheep without being trampled upon or dirtied. They are indeed a most excellent crop for cultivation on those heavy soils unsuited for the growth of turnips, and are more nutritious and better liked by sheep than common turnips. The green or purple small-topped varieties are the best.

The kohl-rabi, as well as the cabbage and the kale, may either be drilled exactly as swedes, at the rate of 2 lb. or 3 lb. per acre, or a few pounds of seed may be sown in a special bed, and the young plants transplanted as required. Provided the weather is not too dry, far from being injured, the plants seem to thrive better for their removal, and the value of these plants, especially the rabi, for filling up gaps in other crops cannot be overestimated.

I have found it a very good plan to drill an acre or two of rabi or cabbage, and then to pull up the surplus plants for transplanting elsewhere. The ground should be damp when transplanting is done, or the plants must be watered, but it is wonderful what they will stand in the way of adverse conditions. I remember once planting several acres of cabbages, straight on to the furrow-slices, after ploughing in a heavy dressing of dung. Two or three weeks of dry weather happening to follow, the plants seemed to disappear entirely, but on rain coming they shot out again, and a good crop was the result. Cabbage provide enormous crops—

from 30 to 40 tons per acre—and should be planted from 18 to 24 inches apart in the row.

Early varieties, if planted out in the autumn, will come in for use in the late spring, and if planted in spring will be ready for summer use. The drumhead cabbages provide food in early winter, and savoys may be used right through the winter. Cabbages are much relished by stock in the yards, and are simply invaluable for carting into the fields for stock, when the grass becomes scarce in August and September, or may be consumed by sheep on the land.

Although thousand-headed kale may be fed to all kinds of stock, it is usually grown for sheep feed, and nothing can quite replace it for providing feed in early spring. If well manured it should become, by the end of April, a dense mass of foliage as high as the sheeps' backs. Sown in succession in the spring and early summer, it provides a succession of autumn food, and grows a great weight per acre. Thousand-heads, although they may be transplanted, are usually drilled where they are to grow, and are roughly singled out to about 12 inches apart in the rows. All these plants are gross feeders, and require large quantities of nitrogen. The culture and manuring are therefore exactly the same as that detailed for mangolds, either with or without dung. When cabbage and kale are sown in the autumn for spring food, the top-dressings of nitrate of soda should be applied in February and March. I have one warning to give. The extensive root-system of these plants allows them to utilise the food of the soil to the utmost, so that when they are removed entirely from the land they leave it in very poor condition for the following crop. This may be remedied by the consumption of at least a part of the crop on the land with sheep. Cabbages are said to keep if properly pitted, but I have found this plan an entire failure. Although they may be grown on any land, these crops are more particularly suited for the heavier classes of soil.

The culture of maize as a forage plant may be mentioned here, for its uses are very similar to the crops we have just spoken of. Many experiments have been tried to find some variety better suited than others for culture in this country, but so far no definite results have been reached. I have seen excellent pieces of maize grown by cow-keepers in Essex from ordinary commercial seed. Twenty tons, or even more, of green food can be grown per acre, but the land must be well manured as for cabbages or mangolds. The seed must be deeply drilled in May, at the rate of 1½ bushels per acre, in rows about 18 inches apart. The greatest care must be taken to well cover the seed, and to protect it against rooks, or they will ruin the whole crop.

The after-cultivation consists of two or three horse-hoeings, and the crop arrives at its most nutritious stage some little time after the cobs are well formed. I have seen it carted on to the pastures for cattle, but this method entails great waste, as the thick stalks are not consumed; it is much better cut up with a chaff-cutter. Its nutritive value is about twice that of turnips; it is relished by all kinds of stock, and when cut into chaff it makes first-class silage. The greatest weight per acre can undoubtedly be raised by singling the plants out to about 12 inches apart in the row, and I have seen crops thus grown more than 8 feet high, with stalks as thick as one's wrist; but I contend that, for farm use, a larger number of small plants is far better, as they are more easily handled, and not so coarse.

Where a large quantity of green food is required in late summer, maize is an excellent crop, and for dairy cows is much superior to cabbages, which have a tendency to taint the milk. It must be remembered, however, that all these crops require heavy manuring and a good deal of labour, whereas tares and some other leguminous plants require only a little super-phosphate, need no hoeing, and although the produce may not be so great, still it is more nutritious, and the land is improved in condition by the nitrogen left by the crop.

Rape is a crop largely grown for sheep, and very rapidly produces large quantities of green food. It is usually drilled in rows somewhat narrower than for turnips, at the rate of about 6 lb. of seed per acre, and is afterwards roughly singled out to 6 or 8 inches between the plants. As a catch crop it may be sown broadcast and lightly harrowed in, when of course no hoeing or singling is required. When a large crop is required it should be manured; about half the dressing recommended for mangolds would be very suitable.

Mustard is principally grown as a cover crop or for ploughing in. Sheep will eat it, but it cannot be recommended for feed, and the sheep should never be kept on it for long. Black mustard is the variety grown for the manufacture of table mustard, but the cultivation of this crop is so limited that I shall not enter into it here. White mustard is the kind planted for use while green, and it should be sown broadcast at the rate of from 14 to 20 lb. per acre. It may be sown at any time from June to the end of August, the ground being ploughed and harrowed, or it may be harrowed in or drilled on the stubble. It is not manured as a rule, but if cover is badly wanted it may be given a little nitrate of soda.

Buck-wheat or brank is another cover crop often grown. It may either be fed off or ploughed in ; or it may be allowed to ripen its seed for harvesting or as food for game. I know of no crop more attractive to game than this, but my experience is that they do not lie nearly so well in it as in mustard. For harvesting it should be sown in May, or, for cover, in June or early July. One bushel per acre is sufficient if drilled in rows 12 inches apart, or 1½ bushels may be sown broadcast and harrowed in.

Lupins are not grown nearly so often as they might be. They will thrive upon soils so sandy and dry as to be almost barren. For these soils their value is incalculable, and it is very striking to see a luxuriant crop from 2 to 3 feet high flourishing under such conditions. Being a leguminous plant, the lupin is a nitrogen-collector, and one of the most active, so that the soil is considerably enriched by its growth. For the improvement of barren soils I should strongly recommend the growth of repeated crops of lupins, and if these were manured with small dressings of superphosphate and kainite and the produce ploughed in, the enrichment of the soil would progress rapidly. Many thousands of acres of sandy waste in Germany have been brought into cultivation by this means. Lupins should be drilled in rows 15 or 18 inches apart at the rate of about 1½ bushels per acre, after ploughing and harrowing, and should be rolled immediately. Sheep will eat them to some extent, and they make ideal game cover; but if they are grown for the sake of the land, they must be ploughed in when beginning to flower.

With the exception of the sugar-beet, all the crops mentioned in this chapter are intended for use on the farm, either in improving the soil or to provide a succession of food for live-stock.

I have hesitated as to whether I ought to introduce the potato at all, as it belongs rather to market gardening than to farming, but is now so common a crop, and adds so largely to the sales of the farm, that I must say a little about it. My remarks, however, must be very brief, and confined to main-crop or late varieties.

The land should be thoroughly and deeply cultivated, and then thrown into ridges or drills with the double-breasted plough, the manure being placed in the furrows so formed. Good crops of potatoes can be grown with either farmyard manure alone or with artificial manures alone, but the most economical dressing is a mixture of both. About 12 loads of dung per acre should be spread in the furrows, and a mixture of 3 cwts. of superphosphate, 2 cwts. of kainite, and 1 cwt. of sulphate of ammonia should then be distributed broadcast over the whole. Twice these quantities should be used, if the dung or artificial manure is used alone. As soon as the manure is spread the seed-potatoes should be dropped at intervals of from 12 to 18 inches on to the top of the manure, and the ridges should be immediately closed with the plough. When this work is finished, run a set of harrows over the ridges so as to reduce the depth of earth covering the seed.

As soon as the weeds begin to appear, begin to use the horse-hoe, and continue to do so at frequent intervals, till the potato-tops begin to get bushy. The best horse-hoe for work between ridges is the Planet Junior, for it can be used to stir the ground thoroughly to a depth of several inches, or the sweeps may be set to skim the weeds from the sides of the ridges; or turned in the opposite direction, these sweeps will throw the soil on to the ridges and so earth up the potatoes. This earthing up is of the greatest importance, and should be performed earlier than is usually done, for it not only provides an earth mulch which keeps in the moisture, but seems to prevent the tubers from becoming diseased.

The spraying of the potato-tops in July with Bordeaux mixture is productive of great benefit. Not only does it largely prevent disease, but it also prolongs the life of the tops, and, as the tops have to manufacture the starch for the tubers, the longer these tops are green and active the larger will be the crop. There are several spraying machines on the market, and full directions as to the making and use of the mixture can be obtained with them.

Potatoes are too valuable, as a rule, to feed to stock, but there is always a percentage of unsaleable tubers, which are excellent food for either cattle or pigs. By the method I have indicated it is possible to grow from 10 to 12 tons per acre upon most fairly good soils, so that even at moderate prices the return should be remunerative.

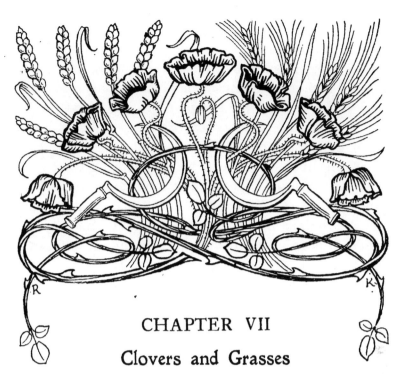

CHAPTER VII

Clovers and Grasses

DURING the last twenty-five years something like three million acres have been sown down to permanent pasture, and at the first glance it would seem rather strange that there has been no corresponding increase in the numbers of our live-stock. If we examine into the matter we can see it is only what we might have expected, for a very large percentage of these acres have been land of the poorer description, which required careful and constant tillage and manuring to be kept in a state of fertility. This it has never received, for it has been considered sufficient in most cases to sow a few grass seeds, and then to allow the resulting herbage to struggle unassisted with the adverse conditions, and even to be repeatedly robbed by mowing without any manurial compensation whatever. It is not surprising, though very reprehensible, that we should in this country have hundreds of thousands of acres only capable of carrying a half-starved sheep or two to the acre.

Although a certain acreage of good permanent pasture is of great value on most farms, it must be remembered that the quantity of food for cattle is never so great as that produced by that same land under the plough, and that the return of wealth to the country is probably much less than half that produced by the same acreage under tillage. The enormous agricultural production of Denmark and consequent increase of wealth and entire absence of depression amongst her farmers, are largely due to the fact that their dairy and other cattle are entirely maintained on the produce of arable land by a succession of temporary pastures and forage crops.

I have already mentioned a considerable number of forage crops capable of providing succulent food for cattle and sheep all the year round, but these have required cultural operations and are not suitable for grazing. The clovers and grasses provide us with crops requiring no tillage of the land for their cultivation, and crops which can be grazed by any kind of live-stock.

The expression "clovers" includes a number of leguminous plants, some of which are not true clovers, but all of which can be classed together agriculturally. The principal plants belonging to this class are the Broad Red Clover (*Trifolium pratense*), the Perennial Red Clover or Cow Grass (*Trifolium perenne*), the White or Dutch Clover (*Trifolium repens*), the Crimson Clover or Trefolium (*Trifolium incarnatum*), Alsike (*Trifolium hybridum*), Trefoil or Black Medic

(*Medicago lupulina*), Lucerne (*Medicago sativa*), and Sainfoin (*Onobrychis sativa*).

All these plants must be sown the year before they are required for use. The Crimson Clover is drilled upon the stubble of the preceding crop as soon after harvest as possible, and all the others are sown amongst the preceding corn crop in April or May. The seed may be sown broadcast, but in my opinion it is always better drilled. One sweep of a very light set of harrows may be given, but in any case the land should be rolled immediately to give that solidity which is so requisite to the proper growth of clovers. The drilling amongst the young corn does not seem to injure it at all if performed when the land is dry, for, except in the case of sainfoin, the drilling should be very shallow.

All these plants do best on soils which contain abundance of phosphates, potash, and lime, so that where one or more of these ingredients are known to be lacking, the application of small quantities will often produce greatly increased crops. We should never lose sight of the fact that there is a double chance of a profit by the application of mineral manures to leguminous plants, for anything which will assist the plant to more robust growth also results in the storing up of increased quantities of nitrogen. Another point must be borne in mind : many failures to obtain profitable results from the use of manures to these plants is due to the application being made too late. It is my experience that to obtain the best results the manures should be applied to the crop in which the "clovers" are sown, so that the young plants may obtain the benefit from the very beginning. By attention to these details good crops can be grown almost anywhere. The condition

L

known as "clover sickness" may prevent the growth of red clover at intervals of less than eight years or so, but there is nothing to prevent the successful cultivation of alsike, trefoil, sainfoin, and other plants in the interval. The cause of clover sickness, that mysterious dying of the clover plants, is, as yet, unexplained, and until the cause is discovered it is useless to suggest a remedy.

The Broad Red Clover is the most extensively grown, and is suitable for all kinds of land. It is sown at the rate of from 10 to 16 lb. per acre, and comes into flower about the middle of June, when it should be mown for hay or fed off. In from six to eight weeks a second crop will have grown, which, if the land is to be considered, should be either eaten on the ground by sheep or ploughed in. If not allowed to produce its seed Broad Clover will sometimes stand a second year, but the crop is never satisfactory.

The Perennial Red Clover stands a second season better, but it is not often sown alone, being usually mixed with grasses, for standing two or more years. It is treated in exactly the same way as Broad Clover, over which it possesses no great advantages for ordinary purposes, although it is two or three weeks later in flowering.

The Crimson Clover has some uses entirely its own Being drilled on the stubble directly after harvest, at the rate of about 20 lb. per acre, it gives us an opportunity of filling up any failures in the plant of other clovers, or of sowing another field or two should it seem necessary. It matures very early, producing its elongated and brilliant crimson heads by the end of May, when it is ready to mow or feed off. An excellent practice, which I have seen carried out very successfully on the heavier soils, is to sow Crimson Clover on a field which requires a little cleaning, and then by the end of May the crop can be consumed or mown, so as to give us the whole of the summer in which to clean the land.

Trefoil is another crop which is ready for use very early in the spring and is often grown previous to fallowing. It produces excellent early sheep feed when mixed with Italian rye grass. When sown alone, from 14 to 20 lb. of milled seed or 1 bushel of the seed in its black husk should be drilled in the corn crop, or half the seed can be replaced by 12 or 14 lb. of Italian rye grass. Trefoil grows well upon the heaviest clays, and makes excellent hay if cut fairly early.

Alsike is an exceedingly useful plant which should be included in all mixtures of clover and grasses, and as it does not suffer from sickness, it may be used in place of red clover, either alone or mixed with trefoil. For sowing alone, 8 lb. per acre should be ample, and it may be expected to produce nearly as much weight as red clover on suitable soils.

The White or Dutch Clover is the clover which predominates in all the best pastures, and, being a true perennial, should be included in the seeding of all pastures intended to remain for two or more years. No plant is so true an index of the supply of mineral food in the soil as this, for it refuses to flourish in soils deficient in phosphates, potash, and lime, and the application of these substances to pastures often produces effects upon the white clover which seem little short of miraculous. I have seen a poor grass field upon which clover seemed non-existent changed in one season by a dressing of basic slag into a dense turf of white clover covered with the sweet-smelling blooms and alive with the hum of bees. White clover is practically never sown alone for a crop, as its produce is small during the first season, but 2 or 3 lb. of seed per acre are included in all mixtures.

Sainfoin is becoming more widely recognised as a valuable plant for certain classes of poor soil, but it will only flourish where the soil or subsoil contains abundance of lime. An exceedingly deep-rooted perennial, it seems able to obtain food and moisture in positions practically barren to other farm crops. Sainfoin is a nitrogen-gatherer, and poor soils are considerably improved by now and then eating this crop on the land. The seed, which is usually obtained in the prickly husk, should be drilled about 1½ inches deep at the rate of 4 bushels per acre in April. Although clovers are commonly drilled amongst barley or oats, sainfoin should if possible be sown amongst wheat, for the extra solidity of the ground seems to favour its germination and the establishment of the plant. The first crop is often a small one, but during the next four or five years crops of from 1½ to 2½ tons of hay per acre may be expected. There are two varieties of sainfoin—the Common and the Giant; this latter being the better, as it produces two crops each season. I have seen 2 tons of hay per acre mown by the last week in May and a second crop of about 1 ton secured in July. There is no doubt that sainfoin should occasionally be sheeped off, especially the second crops, and I know of no green food which will produce larger and finer lambs than this.

Although a plant of sainfoin will last from 8 to 10

years, yet it is inadvisable to allow it to remain so long, as the land gradually becomes covered with weeds which are difficult afterwards to eradicate.

Lucerne is another perennial crop which is open to the same objection in this respect as sainfoin. It is, however, a valuable crop, and every farmer should, in my opinion, possess, at any rate, a small patch. Sown in exactly the same way as clover, it will last for several years, and will provide three and four cuttings a year. It is difficult to make into good hay, as the leaf drops off so easily when dry; but when cut just before it blooms, it will provide 15 tons per acre, or more, of excellent green fodder for horses or cattle, and a good cut can be obtained by the first week in May. It often pays to manure the lucerne patch with phosphates and potash, and even a little nitrate of soda in the early spring; but if dung is used, care must be taken that it does not lie on the crowns of the plants, or it will cause them to rot.

In many places where lucerne is sown for the first time it is difficult to obtain a plant owing to the seedlings dying off. This has been remedied by a load or two of soil from a lucerne field in the neighbourhood, and which may either be drilled with the seed seed or sprinkled over the ground and harrowed in. The cause of this is very interesting. Professor Nobbe, when investigating the nitrogen-fixing bacteria which live in the roots of various leguminous plants, found that each species of plant had its own particular bacterium, without which it could not thrive. If these were present in the soil the plant grew well; if not, it died. He therefore prepared pure cultures of the bacteria of the agricultural leguminous plants, which he placed on the market in bottles, under the name of "Nitragin," and the particular bacterium for any leguminous plant can thus be obtained. Experiments have shown that our British soils are usually well provided with all the suitable bacteria, except sometimes those of lucerne. In this case we can either inoculate the soil by means of nitragin, or by some soil which we know contains the proper bacterium of the plant we require.

Much more might be said about this exceedingly useful group of plants, but I think I have said enough to show how various may be their uses in the economy of the farm.

The clovers are often sown in mixture with various grasses, either to be mown or grazed for one or more years, and are then broken up again. The practice is a very good one, especially when grazing takes place, for although we can cut large crops for hay, it must be remembered that, when mown, grasses largely neutralise the fertilising effect of the clovers.

When suitable mixtures are sown, a much greater amount of grazing can be obtained by this method than from most permanent pastures, and at the same time our land is manured by the cattle. These temporary pastures can be sown to last as long as four years, but I think the two-year plan is the best. The first crop is mown for hay, and then the field is grazed for the remainder of the first year and all the second year, after which it is ploughed up, to be followed by wheat or oats.

On land in fairly good condition it should not be necessary to manure for the hay crop, but where the land is poor, or as much grass as possible is required, manures may be profitably employed. Dung is often applied as a top-dressing on the young plants during the winter, but I believe that the dung can be more profitably employed elsewhere, and that the most reliable results can be obtained from artificial manures. Phosphates, and potash, if required, should be applied when the seed is sown; and $1\frac{1}{2}$ cwt. of nitrate of soda or 1 cwt. of sulphate of ammonia, given in early spring before mowing, will greatly help the grasses. Nitrogenous manures are often said to make the grass washy and of poor feeding quality; but, although there might be some danger of this if we used excessive quantities of nitrogen without minerals, there need be no fear of any loss of quality when manuring is done as I have advised.

The grasses most suitable for mixing with clovers for the production of temporary pastures are, Perennial Rye grass (*Lolium perenne*), Italian Rye grass (*Lolium Italicum*), Cocksfoot (*Dactylis glomerata*), Timothy (*Phleum pratense*), Meadow Fescue (*Festuca pratensis*), and Hard Fescue (*Festuca duriuscula*).

All these grasses will grow upon any soil, provided it contains sufficient moisture and available plant food. It must be remembered that the grasses belong to the same order of plants as wheat, barley, and oats, and that the rules of manuring given for these crops apply equally to the grasses; nitrogen will give us quantity, but a proper balance of mineral manure is required for quality.

All grasses prefer a rich and fairly moist soil, and will grow with the greatest luxuriance in such a situation. I have often seen it stated that such and such a grass thrives upon poor soils, but no grass thrives nearly so well upon a poor soil as it would upon a rich one—it merely manages to grow there because it can put up with greater vicissitudes than some others. It is a case of the survival of the fittest. The wiry little Sheeps' Fescue grass manages to cover the thin soil of the mountain-top because it can exist in spite of an occasional dryness which kills its larger-growing rivals; whereas in the rich meadow this little fescue would be smothered by the Foxtail and similar grasses. The Sheeps' Fescue may not attain to more than 2 or 3 inches in height on the mountain, but I have seen it reach 18 inches in a rich, moist soil where it was free from competition. This keen competition for food and moisture, and more particularly for light, is the factor that determines which grasses shall grow and which shall die, and the plant which can make the best use of the prevailing conditions wins, crowding out its less fortunate rivals. It is almost saddening to think that our peaceful meadows are the arena of this never-ending struggle for existence, this fight to the death. We, as farmers, cannot afford to look unconcernedly on this struggle, for we require those plants to succeed which are of the greatest use to us and our cattle, and we can, by the use of manures, or by drainage or irrigation, completely change its results.

By manuring with nitrate of soda we can encourage the large-growing, deeply-rooted grasses to the detriment of the smaller kinds; but sulphate of ammonia, not being so readily washed downward, will favour the growth of the finer and shallow-rooted kinds, which, seizing the nitrogen first, starve the others. With mineral manures alone the grasses all remain stunted, and those leguminous plants which can thrive without nitrogen in the soil usurp their places. Drainage means the crowding out of the Creeping-bent grass, Yorkshire Fog, and other plants which can thrive in damp situations, while occasional irrigation encourages Rye grass, Meadow Foxtail, and Meadow Barley grass. In each case it merely means that some plants found themselves able to crowd out others by depriving them of food and light, and not that those crowded-out plants would not have succeeded had they had the space to themselves.

All plants grow best upon a rich, well-tilled soil, but it is no use the farmer planting those grasses which he knows will soon be smothered by others more robust than themselves.

The grasses recommended for temporary pastures are all of the robust type, and attain greatest perfection upon a good, well-manured soil. Italian Rye grass, the largest of these, is much relished by stock, and produces an enormous weight per acre; but being a biennial, it is only of use for one year. When mown for hay the cutting should take place directly the flower begins to open, for if left later the hay is apt to be hard and strawy.

Perennial Rye grass is only really perennial on rich soils, and, in my opinion, it is one of the most valuable of our grasses, in spite of all that has been said against it by Mr. De Laune and others. It forms a large proportion of the herbage of our best pastures, it produces a large crop, it is readily eaten by cattle, and it starts its growth very early in the spring; but one might as well try to grow roses in a sand-pit as rye grass to perfection on land that is poor.

Cocksfoot is a very good perennial grass, which is

able to maintain itself in droughty seasons owing to its depth of root. It starts growth early in the spring, and, although very coarse-looking, is readily eaten by stock. It must be grazed or mown early, or it becomes hard and innutritious.

Timothy grass is a tall, coarse-growing perennial, but is excellent for including in mixtures, for grazing especially, as, its period of growth being later than most others, it provides a succession of young grass. It attains its greatest growth upon rather damp, black, clay soils, and I remember, a few years ago, gathering a large sheaf of this grass, every stalk of which was over 6 feet in height. Hay made from this grass alone is largely used in America, but it is very coarse, and fortunately, when grown in mixtures, it is always cut before it has reached maturity.

Meadow Fescue and Hard Fescue are excellent grasses for mixtures intended to stand two or more years, but they do not make much show amongst the others in the first season. Their function is to provide bottom grass for grazing purposes, and although both may be included in any mixture, yet the Meadow Fescue will succeed better on the heavier, and the Hard Fescue on the lighter soils.

It is not a very difficult matter to select from the various clovers and grasses I have named, the varieties likely to succeed in most situations.

For one year's mowing it is commonly the custom to sow a mixture of about 14 lb. of Italian Rye grass and 8 lb. of Red Clover; or where clover sickness is prevalent, 4 lb. of Alsike, 4 lb. of Trefoil, and 14 lb. of Italian Rye grass. Where there is a probability of grazing, a more complex mixture may be used—say, Italian Rye grass,

5 lb.; Perennial Rye grass, 4 lb.; Cocksfoot, 4 lb.; Timothy, 2 lb.; Red Clover, 4 lb.; Alsike, 1 lb.; Trefoil, 1 lb.

When the pasture is likely to be grazed a second or third year, the following mixture may be recommended: Italian Rye grass, 4 lb.; Perennial Rye grass, 4 lb.; Cocksfoot, 5 lb.; Meadow Fescue, 3 lb.; Timothy, 2 lb.; Perennial Red clover, 3 lb.; Alsike, 2 lb.; White clover, 2 lb. All these quantities are, of course, per acre.

Although Rye grass seed can be drilled in exactly the same way as clover, it is generally best to sow grass seeds on the surface, and to cover them in with a light set of harrows, to be followed by the roller. The different weights of the individual seeds in a mixture make them difficult to broadcast by hand; the proper seed-barrow is undoubtedly the best implement, but they can be regularly sown with a common drill, care being taken to fasten up the coulters so that they miss the ground by 3 or 4 inches.

A very fine surface tilth is essential to the proper germination of grass seeds, which cannot grow amongst hard and dry clots. April and the beginning of May is the best time for sowing, and care must be taken, if the land is in good condition, not to sow the corn crop too thickly, or there will be some likelihood of the young grasses and clovers being smothered. In sowing the quantity of seed which experience has shown is necessary, we provide a very large margin for waste by non-germination or death, for in the last mixture I gave there are nearly 11 millions of seeds, or about one and a half per square inch of surface. In the case of permanent pastures it is usual to sow from 18 to 20 millions, as the death-rate amongst the smaller and more delicate plants is so enormous.

For permanent pastures some other grasses are added to those I have already named. Meadow Foxtail (*Alopecurus pratensis*) is the earliest of all our British pasture grasses. It does not show much for a year or two after sowing, but when once established it produces a large yield of excellent quality.

The Rough-stalked Meadow grass (*Poa trivialis*), although a fine and comparatively small grass, can hold its own in rich, damp soils. It is much relished by stock, and, owing to the large amount of leaf it produces, is valuable as a bottom grass.

The Smooth-stalked Meadow grass (*Poa pratensis*) is very similar to the Rough-stalked Meadow grass in appearance, though not so valuable, but it survives well in dry situations, and has a value for sowing in these places.

Crested Dogstail (*Cynosurus cristatus*) is often condemned because the stock refuse to eat the flower-stalks when they become hard and dry. As, however, it produces a dense mat of leaves, of which sheep are very fond, I believe it is worthy of a place in our mixtures intended for heavy clays, for it quickly covers the ground, and if larger and stronger grass can grow, it will soon be crowded out.

The Yellow Oat grass (*Avena flavescens*), when in flower, is perhaps one of our most beautiful grasses. It produces a fairly large amount of very fine herbage, but it only succeeds on a chalky or marly subsoil; in fact its presence in natural pastures may be taken as an indication of the abundance of lime in the soil.

The Tall Oat grass (*Avena elatior*) is sometimes

sown, but it is a coarse and bitter grass, disliked by stock. Its only recommendation is that it will grow upon soils too dry and barren to bear anything else.

The Tall Fescue (*Festuca elatior*) is a large and coarse grass sometimes recommended. It possesses no advantages over the Meadow Fescue, its greater size being counterbalanced by the unpalatableness of its leaves and stalks.

The seed sold as that of Fiorin grass (*Agrostis stolonifera*) should never be sown—it is a weed.

Several other plants which are neither grasses nor clovers are often included with the seeds for permanent pastures. In my opinion they are weeds, and should be rigorously excluded. I allude to Chicory, Burnet, Sheeps' Parsley, and Yarrow. They are said to be beneficial because of their deep roots, but Sainfoin and Lucerne are quite as deep rooted, and do not smother out the grasses by their spreading leaves. There are soils and situations upon which it is difficult to get the good grasses to grow, but until we have exhausted our long list of useful plants, and particularly of leguminous plants, let us beware of sowing anything which may become troublesome as a weed.

I have sown down, and have seen sown down to permanent pasture, some hundreds of acres, and I know that it is very difficult to say that any particular mixture of seeds is the best for any kind of soil; we can only sow what is apparently the most likely to succeed, for the numbers of the various grasses which survive seem to bear no proportion to number of seeds of the different kinds sown.

I should recommend such a mixture as the following for heavy soils :—

Meadow Foxtail	3 lb.
Cocksfoot	6 „
Meadow Fescue	6 „
Perennial Rye Grass	7 „
Rough-stalked Meadow Grass . . .	1 „
Hard Fescue	1 „
Crested Dogstail	1 „
Timothy	4 „
Alsike	2 „
White Clover	2 „
Perennial Red Clover	3 „
	36 lb.

Farmers are sometimes recommended to replace 2 or 3 lb. of Perennial Rye grass by the same quantity of Italian Rye grass for the sake of the larger cut obtained the first year.

For light soils, replace 2 lb. of the Meadow Foxtail by the same quantity of Hard Fescue, and change the Rough-stalked to the Smooth-stalked Meadow grass. On sandy soils 2 lb. of Lucerne may be added with advantage, and on chalky soils half a pound of Yellow Oat grass should be put in. What our pasture may turn out to be will depend more on how we treat it than on the presence or absence of a pound or two of any particular seed, but I must add a few words of warning as to the purchase of seed.

Grass seeds are often very impure, sometimes from accident, but, I am sorry to say, very often from intentional adulteration. Seed worth 4d. a pound can be added to seeds worth 1s. or more without any fear of detection by the ordinary farmer, and seeds of a worthless weed grass, like the Black bent, are added to the expensive seeds of Meadow Foxtail.

Again, the percentage of germination varies greatly according to the condition in which the seed was harvested, and it is useless for us to give several shillings a pound for husks which contain no seed. The seed of Meadow Foxtail can be obtained having a germination of 80 per cent., but samples are often sold which will not germinate more than 10 per cent. Seeds of one grass are sometimes sold for those of another kind altogether. The following rules should be observed in purchasing grass and clover seeds :—

1st. Require from the seed merchant, before purchasing, a guarantee of the purity (freedom from admixture of other seeds), of the genuineness (truthfulness to name), and of the percentage of germination of each species of clover and grass seed.

2nd. Require each species of seed to be packed separately, and on no account buy seeds already mixed.

3rd. Take samples from the bulks as soon as delivered, and submit them to an experienced botanist for report, and do not sow them till the report is received.

One would think that no respectable firm would object to warrant their wares, but it is sometimes difficult to obtain a satisfactory guarantee. My advice is, do not deal with a firm that will not give it. I have bought over 4 tons of various grass seeds from one firm on the conditions I have named, and in no case have they fallen below the guarantee.

The greatest care must be taken to mix the seeds thoroughly before sowing, and they should be sown as recommended in the case of temporary pastures. It is not nearly so difficult to make a good pasture as some would suppose, but the mistake is often made of sowing down the poorest fields on the farm, and then expecting it to turn into pasture without any care and trouble. A good pasture is a valuable asset, and is worth some trouble. Remember that to grow good grass requires the same attention to the rules which govern fertility as we give to our corn and root crops, and that it will pay for it quite as well.

Assuming that the crop in which the grass seeds were sown received a liberal dressing of the necessary mineral manures, our young seeds will require no attention till the following spring, and on no account should stock be allowed on them during the first winter, however strong they may look.

As soon as dry enough in the spring, pass over them the heaviest roll you have. If the grasses appear at all weakly give them a small dressing—say, 1 cwt. of nitrate of soda if heavy soil, or sulphate of ammonia if light. Mow early, as soon as the bulk of the flower-stalks are developed. Allow the grasses long enough to grow a little after mowing, and then stock the field with young cattle and a very few sheep, if any. Remove the cattle if the land becomes soft with rain, and do not stock at all during the winter.

The following spring roll again; do not harrow, but spread droppings and mole-hills by hand, and stock the land sufficiently with cattle and sheep to prevent any of the grasses from running to seed.

The end of this second season is the time when it pays to manure all new pastures, and I know nothing better than a fair top-dressing of good farmyard manure applied in early winter. At the end of the third season a good dressing of phosphates, basic slag or superphosphate, according to the soil, will stimulate the clovers, and safely tide the pasture over its most critical period.

It is neglect during the third and fourth seasons that ruins so many new pastures. If dung cannot be spared, or the pasture is far from home, give a complete artificial dressing of phosphates, potash, and nitrogen. A pasture cannot maintain itself, even if grazed, for the first few years. Assist it all you can by manure, and by feeding corn and roots to the stock.

There are many thousands of acres of pasture in this country which have been sadly neglected, but can be astonishingly improved with care. They are poor because they are starved in every case. Want of drainage may be a contributory cause in many cases, because we have seen that stagnant water prevents the soil food from becoming available, and that cause should be removed; but it is food that all poor pastures require, and especially mineral food.

If we look over the districts where all our best pastures are situated, pastures which will fatten a bullock without any cake, we shall find they are growing upon soils rich in phosphates, potash, and lime. Nitrogen they must have too, of course, to produce a quantity of herbage; but I am firmly convinced that it is to their richness in mineral plant food they owe their quality. When we remember the immense quantity of phosphate of lime used by our growing animals in building up their frames, it is not surprising that a pasture should require its stock replenished, and there is no doubt that the first step towards assisting a poor field is to give it a liberal dressing of phosphates. Sandy soils may require potash, too, and a nitrogenous manure may follow if the quantity of herbage is still small, but it is phosphates first.

On one of my farms was a poor clayey field which had been sown down some twenty years ago, and

M

since neglected. It had become studded with stunted thorn bushes, and only grew a little thin wiry grass, practically useless for anything but game cover.

As it was just such a field as the Cambridge University Agricultural Department required for an experiment on the improvement of pasture, it was taken in hand for experiment. During the winter of 1899 the bushes were grubbed up and the field was divided in five plots of 3 acres each by fences of strong wire netting. Early in January 1900 one of these plots was manured with basic slag at the rate of 10 cwts. per acre, another with 5 cwts. of basic slag per acre, and a third with 7 cwts. of superphosphate, which supplied the same quantity of phosphoric acid per acre as the 5 cwts. of slag; and two plots were left unmanured. In the following May a number of cross-bred yearling sheep were fasted, carefully sorted and weighed, and six of these sheep were placed on each plot. The sheep on one of the plots which had received no manure were given about 1 lb. of linseed cake per day per head, but on no other plot did the sheep receive anything but that which they could graze for themselves. The sheep were weighed each month, and it was soon evident that the manured plots required more sheep to keep down the herbage, which had begun to grow in a remarkable manner. Other sheep from the same batch, which had been held in reserve, were therefore weighed on to the plots that required them.

About the twelfth week it was evident that a number of the sheep were fit to kill, and so a butcher was asked to come and select those he thought good enough. He knew nothing about the experiment, but he selected fifteen, four from superphosphate plot, three from the plot getting 5 cwts. slag, five from the plot getting 10 cwts. slag, one from the unmanured plot, and, to the immense disappointment of the shepherd, only two from those receiving cake.[1] At the beginning of the experiment my shepherd was confident of the success of the cake, and it was amusing to watch the disappointment depicted in his face as week by week he saw these sheep slowly but surely being beaten by those grazing the slagged plot next to them.

In the following two years the proceeding has been similar, except that smaller sheep have been used, and they have been allowed to remain unchanged throughout the twenty weeks during which the experiment lasted. The following table gives a summary of the results for the years 1900, 1901, and 1902 :—

Plot.	Treatment per Acre.			Average Number of Sheep per Acre carried for 20 Weeks each Year.	Live-Weight Gain per Acre.	Average Gain per Sheep per Week.	Total Live-Weight Gain per Acre during the 3 Years.
					lb.	lb. oz.	lb.
1	No manure	Sheep received	Cake 1900	2.0	94	2 6	320
			Cake 1901	2.8	133	2 6	
			No cake 1902	2.1	93	2 3	
2	10 cwts. of basic slag for 1900, nothing since .		1900	2.7	142	2 10	456
			1901	2.8	140	2 8	
			1902	3.4	174	2 10	
3	No manure . .		1900	1.8	74	2 1	201
			1901	1.9	50	1 6	
			1902	1.7	77	2 5	
4	5 cwts. of basic slag for 1900, nothing since .		1900	2.1	117	2 12	392
			1901	2.5	119	2 6	
			1902	2.6	156	2 15	
5	7 cwts. of superphosphate for 1900, nothing since . . .		1900	2.4	127	2 10	372
			1901	2.5	122	2 7	
			1902	2.6	123	2 5	

[1] As showing the pecuniary results of the treatment, it may be mentioned that the 15 sheep paid 18s. 6d. a head for the twelve weeks' keep.

This table should require very little explanation, except, perhaps, as to the reason why the number of sheep per acre is expressed in decimals. We tried to stock each plot with as nearly as possible the number of sheep it might be expected to carry in practice, and so the number increased in the middle of the season and decreased towards the end. For example, in 1902 plot 2 carried 9 sheep the first month, 12 sheep the second, 15 during the third and fourth, and only 5 in the last month. As the plot is a trifle over 3 acres, this works out to an average of 3.4 sheep per acre during the five months. That the sheep receiving 1 lb. of cake per day per head should have been beaten by those grazing the land dressed with slag, and receiving no cake, seems very remarkable at first sight, but the same result has been obtained at other places. The figures for plots 1 and 2 in 1901 show that the sheep on plot 2 made weight faster than the same number of sheep on plot 1, which received cake.

With regard to plots 4 and 5, I may say that although the superphosphate produces the most grass, it does not seem to do the sheep so well as the slag on this land; but it must be remembered that there are soils on which the superphosphate would probably beat the slag.

A glance at the two photographs taken in 1901 after the sheep had been on the plots about ten weeks shows the extraordinary difference in the appearance of the sheep of plots 3 and 4. The sheep were all of the same average weight when put on, and one might almost think from their appearance that the sheep of plot 3 were sinking in weight but for the fact that

the table shows they increased at the rate of 1 lb. 6 oz.
per week per head.   The difference in the herbage is
quite as striking as that of the sheep: it has changed
from thin wiry grass to a mass of clover, and there is
no doubt at all that the manured plots are improving
each year, although it is three years since the manure
was applied.   In 1902 each of the manured plots would
have easily carried a fair-sized bullock in addition to
the sheep, for a good deal of long grass was left at the
end of the season.

Now this experiment, confirmed as it is by several
others, shows that for the improvement of poor pastures
it is phosphates first.   After this, a dressing of dung or
the feeding of cake would increase the nitrogen-supply
and so increase the bulk of herbage.   Except upon those
soils already rich in minerals, the consumption of cake
is a slow and unsatisfactory method of improving
pastures; for cake supplies little but nitrogen, and
nitrogen will never increase the quality of the herbage,
even if it makes it grow more bulk.   Bulk is of course
important, and quality having been catered for, the
pastures will be improved by any kind of nitrogenous
dressing we may give them, farmyard manure, road
scrapings, cleanings of stackyards, or nitrate of soda
and sulphate of ammonia in small quantities.   In a
general way we may assert that new pastures require
nitrogen more and old pastures phosphates, but for some
pastures both nitrogen and phosphates may be required.

Poor pastures are always covered with moss, and
many a time I have been asked how it may be got rid
of.   There is one way, and only one way, to free a
pasture from moss: increase the grasses and clovers,
and the moss and other weeds will disappear.

It is an excellent plan to run a chain harrow over old pastures in the early spring, and all grass fields should be rolled. Never allow droppings to accumulate under the hedges or trees; they should be thrown into heaps and carted on to the bare patches. All droppings about the fields should be spread with a fork once or twice a year, and the thistles should be mown in July, just before they flower. If a pasture is worth having, it is worth taking care of, and these details are scrupulously attended to in all the best grazing districts of England.

Much of the difficulty in making good pastures is due to ignorance and neglect, and to systematic robbery of the grass for the benefit of the ploughed land. It is seldom necessary even on the worst pastures to sow renovating seeds; the clovers and grasses we require are usually there, struggling for existence under adverse circumstances, but judicious manuring and care will bring them to the front. Much more might be said about this interesting and important subject, but I think these few hints, gathered as they have been from experience in farming both poor and first-class pastures, may be of use to those wishing to make two blades of grass grow where one grew before.

The preservation of fodder crops for winter use is nearly as important a subject as how to grow them. The art of hay-making varies so much in practice according to the crop and the climate that it is difficult to be at all definite over it, but one or two points that I have observed may be discussed.

So great is the damage done to hay by rain that rapidity in gathering out of the field is of the utmost importance; but the time required to make hay fit to stack varies very much according to the kind of crop and the kind of land, altogether irrespective of the weather. Hay from rich and highly-manured land requires much more drying than the hay from a poor soil, even supposing the crops to be equal in quantity; and all leguminous crops require much more drying than grasses.

All hay should become slightly warm and sink in the stack, but hay which becomes more than a golden brown is injured in quality. It is rather strange that in the southern half of England hay is preferred which has heated in the stack, while in the north heated hay is considered ruined; indeed I have there seen much hay spoilt in the field for fear of a little heat in the stack. In my opinion neither extreme is correct.

Coarse rye grass and cocksfoot hay may be carried as soon it is dry to the touch and before it has become at all brittle, and the same may be said of permanent grass from poor land. Sainfoin and red clover are fit to carry as soon as the leaves are crisp and the stalks withered; but lucerne, trefoil, tares, and the grass from rich pastures, require to be thoroughly dry in every part before they may be stacked with safety.

The mowing-machine gives us rapidity of cutting; the swath-turner for clovers, and the hay-maker or tedding-machine for grasses, give us rapidity in preparing. Clovers and other brittle-leafed plants must still be put into cocks by hand, but grasses can be swept into rows with the horse-rake, loaded on to the waggons with a hay-loader, and put on to the stack with an elevator. The most modern invention in hay-making is the gigantic sweep, which gathers up the hay just as it has been left by the hay-maker, and is drawn with its load to the stack. Here it is hoisted bodily, by means of pulley and horse, straight on to the stack, where it deposits its load and is let down again empty, or it may be made to tip its load into an elevator which carries the hay up to the top of the stack. These are, indeed, changes from the days of the hand-fork and rake, and some of them may be adopted with advantage. The labour of putting hay into cocks should be avoided whenever possible, although there are showery times when it is very useful; but elaborately to build great cocks in fine weather, and to allow them to remain for days and even weeks in the field, appears to me to be absurd.

Ensiling, or the making of silage, is an important means of preserving crops which is not used so much as it should be. The expensive buildings and elaborate pressing apparatus which used to be recommended did much to prevent its adoption by farmers.

Although I have made excellent silage in stacks, yet I much prefer a silo, for in the silo there is practically no waste, the material can be cut into chaff before putting in, and, most important of all, if it is properly made it requires no pressure. Any building with four fairly-strong brick walls makes a good silo; it may either be built for the purpose, or a brick wall may be run across the end of a barn or other building with smooth walls. I have tried boards, but found they bulged unequally and let the air in, and all we have to do to make good silage is to keep the air out. When air is present, fermentation and heating takes place, and according to the amount of fermentation and heating we get sour or sweet silage. The rules are these: when the material is very wet and tightly pressed we get green sour silage, but if the material is slightly dried or is not pressed we get brown sweet silage. Cattle seem equally fond of either.

As my method seems fairly simple I will explain it, although it may not meet all cases. I used a steam chaff-cutter, fitted with a 12-foot elevator for cutting chaff for my cattle, and this was set so as to throw the chaff over the silo wall. Early in the morning the mowing-machine was sent into the crop, and after three or four rounds had been made the carts started bringing the crop home, when it was passed through the chaff-cutter and shot into the silo. A boy distributed the chaff about and two men kept it constantly trampled down, going more particularly round the wall, pressing it closely down by the bricks. Care was taken not to let the mowing-machine get too far in advance of the carts, especially in the middle of the day, so as to prevent drying as far as possible. After perhaps two days the silo was about three parts full, and operations were suspended for a couple of days to allow the material to settle and to give it a good trampling each day. It was then filled up and given a thorough trampling with all hands. Now comes the important point. Three or four loads of fairly stiff pond mud were brought, and with this the whole of the top of the chaff was carefully plastered about three inches thick, and a perfectly air-tight covering was thus formed. As the silage settles, cracks will appear in the mud, which should be carefully filled up every few days for a week or two. Silage made in this way should be of a pale brown colour and smell something like a navy-cut tobacco. The mud covering may be peeled off, and there should not be an atom of waste. Unchaffed grass or other material can be put into the silo, but I much prefer the chaff, as it mixes so well with any other food. Maize, or any such coarse material, must certainly be cut into chaff to be successful. When ensilage is made in a stack the material is better damp or even wet, so that it will pack together closely and prevent too much heating. I believe ensilage will be found a great help wherever a large head of cattle is kept, because it enables us to increase our supply of winter food by preserving crops not easily made into hay, and it renders us less dependent upon the root crop.

At last I have come to the end of my chapters on crops. I have not said nearly so much about them as I ought to have done, nor so much as I should have liked, but I must not tire the reader nor curtail the space to be devoted to live-stock. You must, however, have crops, and know how to grow them, before you can keep stock. When I first started farming as a very young man I was well aware of the advantages of stock-keeping, but had not sufficient knowledge or experience to know that the first step to successful stock-keeping is to prepare food-crops for the stock to eat. I well remember those two first winters. Long before spring appeared everything was consumed, and I had to buy hay, straw, roots, and cake to get over the last few weeks, and I am certain that that buying killed any possible profit for another year or two at least.

A good many years of experience have taught me that you cannot buy food and make a profit on stock, and although it is often necessary to buy concentrated foods, yet the greater the purchases the less the profit on the stock. This is my excuse for having taken up so much space with the food-crops of the farm.

CHAPTER VIII

Live-stock

ALTHOUGH the keeping of live-stock is generally considered the most profitable part of farming, yet it is not always so; for I have repeatedly heard large, and one would think skilful, farmers declare that their cattle often lose them money. It is well known that horses are very precarious in the matter of profits, pigs fluctuate greatly in value, and one even hears complaints, though not very often, as to sheep not paying; one well-known agricultural writer, quite lately, making a calculation which led him to doubt if his flock really left any profit. One might almost think that the stock-keeper was in the position of the old market-woman who, buying her eggs at two for threeha'pence and selling them at three for twopence, declared that it was the quantity that paid. As a matter of fact stock-keeping does pay, and can be made to pay well if carried out with due consideration of the soil and situation of the farm, the capital at command, and upon lines which are within the skill of the farmer.

This latter is the most important: it embraces many heads; and it is by venturing outside the limit of his skill that I have seen many a young and energetic fellow "come a cropper" before he had had time to "cut his sucking teeth" as a farmer. Let me explain what I mean and give a few hints intended for the young farmer, but which will be confirmed by the older. farmers, I believe, as the fruits of experience. A common source of loss is dealing, especially horse-dealing. Nearly all young men are attracted by horses, and one hears of large sums made in single transactions; but you will find sharper men than your-self, and you will be "done" sooner or later. A certain amount of buying and selling is essential in farming, but dealing is a business and is not within your skill.

A farmer often has to buy a considerable num-ber of cattle or sheep to consume his grass or straw and roots, and even the most experienced farmers find it difficult to buy the right sort of stock at the right price so that it may leave a profit when sold again. It is, in fact, upon this class of stock, whether fed for the butcher or not, that the losses one so commonly hears of are made. It is far better to breed or bring up the stock we require than to risk buying and selling again after a short interval. It requires a kind of natural instinct to become a suc-cessful dealer, but the skill required to breed excellent lambs or to rear capital beasts can soon be acquired.

The breaking and training of nags is an attractive snare to the young farmer. There is something so delightful about the training of a young hunter or a high-stepping hackney, and the recollections of one's own youth are so vivid, that I feel loath to condemn the practice altogether, although it is seldom a source of real profit. The high prices occasionally made of well-trained horses are oft repeated, the many failures are wrapped in oblivion. Few men have the hands or the nerve properly to train a first-class horse; a great amount of time is required, which means neglect of the farm, and often leads to dealing. There is no reason why a young farmer should not train a horse for his own use, have an occasional day with the hounds, and even sell if he gets a good offer; but what I have seen so often in the Midlands leads me to warn the farmer to touch that kind of thing with the greatest caution.

The training of a sound four-year-old cart colt is a different matter, and belongs to the business of the farm. The same farmer who spends hours over a ewe-necked, straight-shouldered hunter allows his much more valuable cart horses to slouch along with their heads between their knees, as if they were ashamed of themselves. It is astonishing how little training it takes to teach a cart colt to hold his head up as if the street belonged to him, and to develop that gay carriage and springy step so attractive to the long-pursed buyer.

The stock of all kinds on the farm should be well bred, the dams should be of the right stamp, and the sires should be pure bred and first-class specimens of their kind. Such a standard of excellence is easily attainable, and is within the skill of any one having enough knowledge to start farming with chances of success. To attempt the breeding of special kinds of stock, or to endeavour to build up pure-bred studs, herds, or flocks, requires a very special knowledge which only comes from experience. The market for some breeds is very limited, and there are others which cannot possibly pay unless specimens can be sold at high prices for breeding purposes, and in both these cases the high standard of excellence required to secure

remunerative prices only comes from an intimate knowledge of the breed and years of experience.

There are, however, breeds amongst horses, cattle,
sheep, and pigs of such commonly recognised utility
and excellence that good specimens are always in
request. I am perfectly certain that the adoption of
some pure breed as a speciality is an excellent practice:
it raises our farming from the common rut, it gives
something in which we can take a keen personal interest,
and with care and judgment it may be made a sound
financial success.

The amount of capital at command is a very important factor, and must considerably influence our choice
in this matter; but to start with a few excellent specimens of any breed of cattle, sheep, or pigs need take
no very great amount, although shires, hackneys, and
even some breeds of cattle require a fairly long purse.

The first considerations when choosing a breed are:
that we must know something about its characteristics;
that it should be in considerable demand either generally or in our particular neighbourhood, and that it is
within our means to obtain good examples. Be content to start in a small way; buy good characteristic
specimens, and avoid fancy prices. Do not attempt
to set the Thames on fire directly you start, and
avoid showing as much as possible. Showing is an
advertisement and nothing else; it is very expensive;
it ruins your breeding stock; and until you have something very good to advertise, confine your efforts in
that direction to a local show or two.

Ruthlessly cull out all inferior animals, and freely
emasculate all males that are not first-class; they will
pay better so. Inferior animals run with the others

spoil the whole lot and make a bad impression on would-be buyers. Your very best breeding stock should never be for sale; they are too valuable to you, and you might not be able to replace them; but introduce new blood repeatedly, taking care you get it from a good source. Choose your males to remedy deficiencies in your dams, and in introducing fresh blood endeavour to obtain it from a flock, herd, or stud that shows strongly the characteristic you are endeavouring to obtain. It is not always sufficient that the individual possesses it, for although like tends to breed like, yet it is those characteristics which seem to be dominant that are most likely to come out.

It is not at all necessary for stock to be pure bred, as horses, cattle, and sheep which make no pretension to purity, but have been well bred to a certain type, are often kept and bred with excellent results. The most prevalent agricultural horse is of the shire type, the commonest cattle of the shorthorn type, and the commonest sheep are crosses between some Longwool and one of the Downs. Stock of this kind, if of a good class, is often quite as profitable as those with the longest pedigrees, but pure-bred sires should always be used in breeding from them.

The particular type of animal which will suit our farm depends largely upon its soil and situation. The texture of the soil will influence our ability to keep a flock of breeding ewes, for it is only upon light and porous soils that a large flock can be kept during the winter. Heavy soils will carry a large head during summer, but as the land becomes wet in winter, very few can be kept round. On very light soils it is often difficult to find summer grazing for many cattle, but with the adoption of temporary pastures more could be kept than is usually the case, and less buying in for the consumption of straw and roots would be required. A fairly porous fertile loam is really the best soil for both cattle and sheep, for if our crops are properly planned, both can be kept in large numbers throughout the year. A heavy soil will grow larger crops and keep more stock in summer than a light one, and as sheep have to be bought in the spring and sold again before winter, such land is more suitable for breeding cattle. Here, with either temporary or permanent pastures, we can keep a large herd, for suitable crops can be grown for summer use, and as winter approaches the cattle are housed to consume our hay, straw, and roots. For the same reasons this class of soil is also particularly suited to dairying.

The quality of the herbage grown upon the pastures of any particular district seems to have a bearing upon the stock, particularly upon the sheep which can be pastured thereon. I have observed that the large-framed sheep, like the Lincolns, Leicesters, or Oxfords, fail to develop properly—without a great deal of hand feeding—upon the thin pastures of the sands and chalk ; while such sheep as the Southdown or Shropshire often fail to thrive upon the luxuriant pastures of the Midlands. Sheep particularly, and cattle to a less extent, have become specialised to a particular habit of life, and to attempt suddenly to change that habit causes failure. The Ayrshire—that magnificent milking breed of the north—is found to degenerate rapidly when bred under southern conditions ; and the Jersey quickly loses its fawn-like appearance and silky skin when allowed the run of rich pastures in England, and becomes nearly as big as a shorthorn in two or three generations. The soil, and the character of the herbage produced, are much more important factors in determining the districts most suitable to our native breeds of stock than the influence of climate, for except in the mountainous districts of Scotland there are no climatic extremes likely to be detrimental to any breed.

The local situation of the farm may influence our choice of the kind of stock considerably. For dairying, and especially for milk-selling, it is essential to be near a town or railway station; whereas in districts remote from these advantages, it becomes important that as much as possible of our corn and other produce should walk to market on the backs of our live-stock.

When fattening for the butcher we also have to study the demands of the market in which we sell. In some markets large and very fat beasts sell well; in others it is only small prime qualities that meet with a good demand. It is just the same with sheep; the large white-faced, the large black-faced, or the small Down, each has districts in which it sells best. It is very unwise to adopt a class which is not fashionable in the district; but as each class includes a number of breeds, we have plenty of scope as to the particular breed or cross we may like to adopt.

The various points that I have mentioned have a very important bearing upon profitable stock-keeping, but having considered them all, and having made up our mind as to the most correct course to adopt, there still remains that all-important subject, feeding. It is a wide experience that the profit on stock is largely determined by the skill and true economy with which

N

the animals are fed from first to last. All farmers
and most labourers know more or less how the various
kinds of stock should be fed, and given plenty of oil-
cake, can make them look well, but the large and
experienced stock-keeper knows that a good deal more
than this is required. It is the skill with which the
proportions of the various home-grown and concentrated foods are adjusted that determines the amount
of animal increase which can be got out of a given
value of food. Long experience has taught many good
farmers how to adjust these proportions fairly well,
but it is possible, by understanding the properties of
the various foods, and the requirements of the animal,
to compound an economical ration without a great deal
of experience.

The subject is well worthy of study; but it will be
sufficient here to state that the food of all animals must
contain certain quantities of the substances containing
nitrogen, commonly called albuminoids, used for the
formation of flesh and other purposes; substances
containing no nitrogen, used by the animal for the
production of heat and energy, the starch, sugar, and
such like substances being spoken of as carbohydrates;
and certain oily matters commonly termed fats. After
the animal has used all it requires for its various life-
processes, it may store up the remainder in its body
as fat. The food also contains mineral matter used
largely for the formation of bone; fibrous substances of
comparatively little use to the animal except to aid the
chewing and digestive processes; and all foods contain
some water.

I shall often have occasion to allude to those substances in dealing with the different kinds of animals;

but what I want to mention here is this: it depends entirely upon the *quantities* and *proportions* of these substances which we give our stock, as to how they will grow, work, or fatten, and, consequently, what profit we shall get. If farmers generally understood the principles of feeding, their stock would be healthier and their profits larger. The albuminoids and fats are the most expensive items of our foods, costing from two and a half to three times as much as the carbohydrates, and yet it is not at all uncommon to see stock of all kinds getting half as much again as they can possibly use. This surplus is not only wasted but causes the other constituents of the food to be wasted too, and sooner or later the animal's health is injured. Young animals, having to grow and make flesh rapidly, require a food containing a fair amount of fat and about one part of albuminoid to four parts of carbohydrate, whereas when full grown they require very little fat and only one part of albuminoid to six or eight of carbohydrate.

Our ordinary farm foods—hay, straw, roots, wheat, barley, and oats—when used in mixtures as food for stock, all contain abundance of carbohydrates, but are deficient in albuminoids and fats, and it is to supply these that we buy expensive cakes. A certain amount of cake, rich in the constituents we require, is often a profitable investment; but the reckless and extravagant use of these expensive foods, such as one often sees, must lead to great loss. If it does not, then the profit on skilful farming must be very great! I know, however, that the most experienced farmers find it difficult to buy cattle, fatten them on bought cake, and sell them at a profit.

The question arises, can we manage to feed our stock within the limit of our own farm ? We undoubtedly can, given certain conditions. It would be possible for us to grow beans and linseed, and by mixing two parts of beans with one of linseed to make a food very little inferior to good linseed cake, though it is doubtful if it would be cheaper. We could even make up a mixture of beans, barley or wheat, and linseed, which, used with hay, straw, and roots, would be quite as effectual as cake, and, unless prices of corn were high, would also be cheaper. Whether it pays to use our own corn or to sell it and buy cakes and other foods entirely depends on their relative prices, and it undoubtedly pays well at times to buy suitable food for blending with those we grow. This matter of the proper use of the concentrated goods is of great importance, and is often a considerable factor in profitable stock-keeping, but it does not go quite to the root of the question.

The farmer who farms on the four-course rotation of roots, barley, hay, wheat, or oats, has practically nothing upon which he can keep his stock during the summer, unless he possesses some permanent pasture or grazes some of his hay land; but during the winter he has to consume produce grown on the whole of his arable land. He can keep very little stock of any kind during the summer by this system, but in winter he must have large numbers of both sheep and cattle, which he has to buy in the autumn and sell again in the spring, thus exposing himself to risk of loss number one ; and as his hay, straw, and roots will not keep or fatten stock alone, he must buy cakes, and thus becomes exposed to risk of loss number two.

It is a well-known saying that two good turnip years following will ruin any four-course farmer; and although, like many other sayings, this must not be taken literally, it shows the tendency to loss to which such a system is exposed. The same kind of thing goes on to a less extent under many other systems of farming. Now, by reducing the area of roots to one-third, the land could be better tilled and a larger crop grown. The remaining two-thirds could be devoted to leguminous and other green crops, to be partly fed on the land by sheep, and partly given to young cattle. Part of the hay area could be grazed and part mown, while the corn crops would remain the same, though it would probably be better to replace some of the wheat by peas or beans.

As the result of this change the farmer could keep a larger flock of breeding ewes, he could breed or rear his own cattle, and the whole of the progeny could be fattened or otherwise disposed of, not compulsorily when the supply of food was finished, but whenever there was the greatest probability of profit. With proper care a larger *average* of stock could be kept, and he would be able to feed them, and fatten them too, almost entirely off his own farm.

It is my contention that there is far more chance of profit in raising a calf costing £2 into a fat beast worth twenty guineas than by fattening two or three beasts costing £15 apiece. For the fattening of store beasts in winter a large allowance of cake or other concentrated food is essential; but cattle, and sheep as well, grow and thrive far better on temporary pastures, clover, tares, cabbages, and other green crops, than they do upon hay, straw, and roots, even with an allowance of corn or cake; and if stock are kept in good thriving condition from their birth, the final fattening is very easily and in-expensively brought about.

In my travels I have come across farmers in various parts of the country who farm upon this system. They breed large numbers of horses, cattle, sheep, and pigs, buying calves at times; they grow a large variety of green crops, planned to come in at all times of the year; they always have something to sell, and, considering the numbers of stock turned out, their expenditure on foods is exceedingly small.

One thing that has always struck me upon such farms is the general air of prosperity pervading the whole establishment. It is, in fact, farming as it should be: the greatest possible return from the least possible expenditure. Why should we have to buy either stock or food when the farm is the proper place to raise them both? The fact of the matter is that we have not yet shaken ourselves free from the old-time system of keeping stock merely as manure-producing machines. We try to farm upon rotations and systems in which this was the only object in keeping stock; we fail to realise thoroughly that times have changed, and that stock-keeping should now be one of the principal objects for which we grow crops.

Again, the farmer who plants a fourth of his whole arable land with roots simply for the purpose of manuring that land, has no object in wishing to grow a maximum crop, for his roots would then be out of pro-portion to the rest of his available foods, and he would be unable to use them all. By making an effort to grow maximum crops on smaller areas, and with a greater variety in the crops so grown, not only will the fertility of the farm be well maintained, but the yields of corn will be quite as large, and the stock kept will be far larger. Kept upon the farm, and fed upon its produce all the year round, the profit of the stock cannot fail, I think, to be greater and more certain, and if at times we require extra food, the amount will be comparatively small.

Stock-keeping is a wide subject; the kinds are many, the purposes are various, and the conditions are often very different. The treatment that I may recommend may not fit in with every case, but I hope the principles will be of general application. It is an old saying that "the master's eye fattens his cattle;" and so it does. Remember that the good stockmaster should know nearly every animal on his farm individually. I have been struck sometimes when going round with large breeders to see how they are able to recall the history and peculiarities of almost every animal, and I know from experience what an interest it gives to the farm.

CHAPTER IX

Farm Horses

THERE are at the present time only three breeds of horses suitable for farm-work—the Shire, the Clydesdale, and the Suffolk. It is quite immaterial which of these three breeds we may select, for all are capable of performing the work of the farm in a satisfactory manner. I do not intend dealing with nag horses, but I may here and there give a hint; for, although they may sometimes form a profitable adjunct, they cannot be regarded as a necessary part of the farm stock. It is not my intention, either, to differentiate between the breeds of cart horses, for that would be much too risky a task. I have been amongst studs of all three, and have heard the praises of each sung by its respective (and respected) owner, and one is perfectly willing to admit that each has points in its favour. My advice is, if you take up a pure breed, take up that which you understand, or at any rate know something about, and think you can most readily sell. If it is only a matter of ordinary work horses, let them be a good stamp of the breed most fashionable in the neighbourhood, and above all avoid half-bred nags of the 'bus-horse type.

Working power is the first consideration in choosing farm horses, but I certainly think that a certain amount of breeding is desirable. I have found it a good plan when breeding to keep two or three more horses than the number actually required to work the farm. For example, if six horses are absolutely necessary, then keep eight, of which four may be breeding mares. All will be available for work during the winter and early spring, with the result that the ploughing and seeding will be well forward, and if two or even three foal, the work of the farm will not suffer.

These mares should be good specimens of their class, and, if possible, sound. I say if possible, because, however desirable it may be that none but sound animals should be used for breeding, there would be very little breeding done if every mare had to have a certificate of soundness. Unsoundness resulting from accident is no detriment, and it does not follow that a foal will necessarily develop a disease from which either of its parents is suffering, though it will undoubtedly inherit a tendency to that disease. Side-bones are the greatest scourge of our cart horses at the present time, and every effort should be made to prevent their recurrence. Misshapen mares should be looked upon with suspicion, for unless we know positively that it was the result of an accident, a bodily defect, such as bent hocks or club feet, is almost sure to be hereditary. Select as a sire a perfect animal of his breed, showing strongly-marked male characteristics, and of good temper. Stallions of feminine appearance are seldom good stock-getters, and a bad temper is very often hereditary from either side. A mare showing a deficiency in any respect should if possible be mated with a sire in which that point is well developed.

Breeding mares are best at work, except of course when suckling, and should never be fat. Fat animals are very uncertain breeders; in fact many an animal is rendered barren for life by being made unnaturally fat for showing. A large allowance of corn is always bad for breeding mares, and is doubly so if of too starchy or of too albuminous a nature. The mares should be out at grass as much as possible, both in late autumn and in early spring. I have found the best way of persuading an uncertain mare to breed is to keep her out at grass all the winter, giving her a shed to run in, a small allowance of oats, and some hay in rough weather.

Keep a sharp lookout against the use of arsenic and other nostrums. They are fatal to breeding, and are much more used than is commonly supposed. An unnaturally sleek and shining coat, not the result of "elbow grease," is suspicious, and when it is coupled with barrenness, the veterinary surgeon should be confidentially consulted. It is hardly necessary to say that a man who uses such things should never be employed, and valuable animals should never be entrusted to the care of a bad-tempered or drunken man.

Extremes of all kinds should be avoided—very hard work or long periods of idleness, too fat or too poor condition, too much coddling or neglect of precaution. No special feeding should be necessary if the ration of the stable is a well-balanced one, but avoid heated or mouldy hay, and any sudden changes of diet.

Opinions are divided about the use of salt, and although I disapprove of mixing salt with the food, I think that in inland districts a lump of rock-salt should be occasionally accessible to all animals. From the beginning of March, 2 or 3 lb. of mangolds a day, pulped and added to the chaff of the evening meal, is an excellent thing for in-foal mares, as it prevents costiveness, and gives a relish to the dry food.

With proper care and management there is really a comparatively small amount of risk with breeding mares. One hears of a great deal of "bad luck," losses of mares and losses of foals, but I am perfectly convinced that bad luck is synonymous with bad management; I have found it so in every department of farming. Losses will occasionally occur, but instead of blaming your luck, I have found it much more effectual to set about trying to find out what could possibly have been the cause of the loss.

Mares approaching the time of foaling should be placed in loose boxes at night, and carefully watched. If everything goes well, there is really no necessity for any one to be present at the foaling; but if anything is wrong, there is no time to be lost, and the veterinary surgeon must be sent for at once. Some few experienced grooms know how to deal with simple mishaps, but a mare is much too valuable to be experimented upon by amateurs.

Half-an-hour after it is born the foal should be up and sucking, and about this time the mare should be allowed to drink about a gallon of water. Now this water should not be the usual tepid stuff known as chilled water, but should be cold. All horses abominate warm water, and although it is better to take the chill off icy water, it should still be cold to the touch. An armful of nice sweet hay and a gallon of crushed oats should be given at the same time.

It is usual to give mares bran mashes, but I consider crushed oats are more sustaining, and, as hot foods are very bad for horses at any time, I do not recommend them. By the end of the first day the mare may return to her usual food, provided wheat-straw chaff forms no part of it.

It is always advisable to get a mare and foal out to grass as soon as possible, and the great drawback to early foals is that they cannot be turned out. A week is quite long enough for the loose box ; if the weather is at all summer-like they should be turned out. Cold rain is the only thing to fear, but at the age of three weeks a foal is really very hardy.

On anything like a fairly good pasture, the mares should require no corn after the middle of May ; and the rapidity with which the foals grow and fatten is remarkable. Mares suckling foals should never be worked, but, if absolutely necessary in harvest, half a day's work may be taken now and then. The most trying period for the foal is when it is weaned about the end of September. A week or two before this takes place the mare and foal should be put in a loose box in the daytime, and fed with hay and a few oats and chaff. The foal soon learns to eat out of the manger, and the mare should be taken away for a time, leaving the foal by itself for increasingly longer periods each day. While the mare is away the foal should be given a few crushed oats, and when it learns to eat them well, and drink out of a pail, the mare may be taken away altogether, and returned to the team without any trouble.

In a week the foal may be turned out again in a field at some distance from the mare, and it should always have a companion—if possible, some other foal or foals. It is exceedingly important at this period to prevent foals from losing flesh, for sinking condition means decreased growth and health. To this end a small allowance of oats may be made—say from 1 to 2 quarts a day, according to the season and quality of the grass. By the end of October the question will have to be answered, Where shall the foal be wintered? Undoubtedly, wherever possible, all young horses are better out at grass— that is, in a fairly good pasture field which has a convenient shelter hovel. An allowance of 2 or 3 lb. of oats, mixed with an equal quantity of chaff, and 6 or 7 lb. of good clover hay each per day, should keep them in capital thriving condition, and they will treat all kinds of weather with equal indifference. No amount of cold seems to hurt colts in the least, but the hovel should protect them from the only thing they object to—cold, wet winds.

This outdoor treatment would be the reverse of economical were the production of meat only concerned, but we want what we cannot get in loose boxes—sound, hardy constitutions, plenty of hair, well-shaped, elastic hoofs, and muscles developed by constant exercise. Given sufficient food, these colts will be far happier than those in yards or boxes. As the grass grows in the spring, the hay may be discontinued, and eventually the oats, if desirable.

If the colts are wintered in loose boxes or yards a similar allowance of food will be necessary, but a little pea or oat straw should be given for them to browse over, and after Christmas a small quantity of pulped roots may be mixed with the chaff. The roots will not only correct any tendency to costiveness, but also help with the coarse fodder to keep the stomach and intestines distended, and this, although it makes a "tubby" colt, will produce a more powerful digestive apparatus and provide a better barrel when the animal is grown up and put on dry feed.

I have often heard it remarked that certain districts do not produce colts with sufficient bone, and that other districts produce the best bone. I may be wrong, but, as far as I have observed, I believe those districts coincide with a deficiency or an abundance of lime in the soil. I have never farmed on land deficient in lime, but I believe the recommendation of Wolff, in his book on farm foods, is well worth a trial. He says that growing animals often have a difficulty in obtaining enough lime for bone formation, and he recommends giving half an ounce of chalk per day in their food. I have seen young bullocks chewing lumps of chalk, probably for this very reason, and it seems to me that foals might be allowed a small quantity.

Exercise is essential to all young animals, and must be provided in some form or other for foals kept in boxes. As soon as there is any grass, turn the animals out, and provide a small allowance of corn, if necessary, till the grass is sufficient.

The yearling colts, if not destined for stallions, should now be castrated, but always allow them quite to have shed their winter coats before operating. Stallions and fillies should of course be kept apart now, but both may lie out all the summer. The following winter I recommend just the same treatment as before, but a somewhat larger allowance of corn may be made. Remember also that our growing animals require a considerable quantity of albuminoid food, so that unless we can supply clover hay, the oats should be supplemented with a few beans, grass hay not balancing well with oats alone.

As two-year-olds, stallions may have to be kept in, but if they are not allowed to see any mares they will usually lie quietly enough together in a well-fenced paddock. They must, however, be got in fairly early next autumn and prepared for their work. The geldings and fillies may be treated as before till they are three years old, when they should be broken in.

This breaking-in should have really begun with the foals, for all young horses should be taught to lead with a halter, and that process cannot commence too early. Once really taught to allow themselves to be handled, they never forget it, and it is then an easy matter to put on their harness and get them used to carrying it. The ease with which this is done contrasts favourably with the scene I have sometimes witnessed: a crowd of terrified colts kept hustled together in a corner by a cordon of shouting men, while one of them endeavours again and again to slip the halter on the required animal. At last it is on, and away rush the frightened animals, while the captured one, with half-a-dozen men hanging on to the rope, dashes and plunges and rears, until at last from sheer exhaustion he stands cowed and trembling. Breaking a colt often stands for breaking the heart of a good horse, but I have seen a properly-handled animal walk off with his harness, take his place in the plough team, and in less than half-an-hour be walking up and down the furrow as if born to it.

The plough or the heavy roller is the best thing with which to teach a colt to work, and with two quiet old stagers to do the pulling and a man on each side to lead, no harm can be done. The men should talk quietly all the time, but no shouting or whip-cracking should be allowed, whatever happens. A couple of hours is plenty for the first day or two, and after that the time can be increased. See that his collar fits well and does not press upon the windpipe. Horses are seldom kickers or jibbers naturally: these faults usually arise from bad breaking.

On this care and management, particularly with regard to the food, depends the health, vigour, strength, and appearance of our teams. Hard work, except perhaps trotting upon the stones, never hurts a horse, and horses which cannot stand the hardest work on a farm without knocking up are badly fed. Teamsmen and grooms, especially the latter, very often have a notion about the requirements of a horse about equal to their knowledge of Greek, although here and there one finds an intelligent man. In certain districts a better rule of practice prevails, but wherever the 'vet' is busy you may be sure of bad practice.

Nature has provided the horse with a very small stomach for an animal of his size, and his natural habit of feeding is to eat and drink all day long. We must therefore try and approximate this by giving food *and water* at as short intervals as circumstances will allow. When a horse has been fasting long he is apt to bolt his food without proper mastication, and as the stomach rapidly fills, some of this imperfectly-masticated food will be passed into the intestines without proper digestion. Again, a heavy drink of water always washes a considerable quantity of food out of the stomach, and if this food has been badly masticated, indigestion and colic are likely to ensue. Investigations show that the first-consumed portions of a feed have been passing out of the stomach for some time before the meal is finished, and so with this rapid passage through the stomach it is essential the food shall be of an easily-digested type. Wheat straw and coarse benty hay are therefore unsuitable foods for horses, and when cut into very fine chaff they often escape proper mastication and become a source of danger. The chaff for horses should therefore be so long that it cannot be swallowed without mastication, and all corn should be crushed to ensure digestion. When whole oats or wheat are fed, a considerable percentage will be found to have escaped both mastication and digestion ; in fact this source supplies most of the food of our town birds.

Although a horse with nothing else to do can support itself upon hay and straw, and may work upon good grass, yet it cannot deal with coarse fodders at a sufficient rate to maintain itself at hard work, and therefore concentrated food must be supplied in suitable quantity. For very hard and for very rapid work it has been found that a horse should not be allowed all the hay he will eat, but a larger amount of concentrated food must be supplied. A much larger quantity of water is also

Q

necessary for rapid work, and horses expected to perform such work efficiently should have water always beside them, so that they may take a little as they require it, and not drink a great deal at once nor go out thirsty. I am convinced that the exhaustion of hunters is largely due to thirst, owing to the ridiculous practice of depriving them of water. Horses are very particular as to the quality of their food, both in taste and freshness, and the food should be of a mixed character and varied slightly to prevent loss of appetite.

A mixture of equal weights of oats and meadow hay contains albuminoids and carbohydrates in almost the exact proportions for working farm horses, and from 27 to 31 lb. of the mixture should be sufficient. This is common experience; indeed, it will be found that the army horses of this country receive 22 lb. of nearly such a mixture; those of Germany 23 to 24 lb., but some straw instead of hay; and in the United States, where the horses appear to be somewhat larger, they receive 26 lb. One finds, too, that the rations allowed by various tramway companies run from 26 lb. to 31 lb., and that the weight of corn somewhat exceeds the weight of hay, which is as it should be for faster work. Farmers seldom make any definite allowance of hay to their horses, but my own experience, and that of other farmers, has shown that from 14 to 16 lb., with 14 lb. of corn, is about the proper allowance on the average.

Experience has, however, taught most horse-keepers that excellent as oats may be, when they form the only corn, they are neither the cheapest nor the best. Horses prefer, and do better on, a mixture of grain of which oats form a part. Two parts of maize to one part of beans forms an excellent mixture, and is

211

more nutritive, weight for weight, than oats. It must be remembered that the husks form from one-quarter to one-third of the weight of oats, and the large white oats having the most husk, at least $4\frac{1}{2}$ lb. in every stone has no more value than straw.

Wheat either boiled or crushed used to be used extensively for horses, and there is no doubt at its present price it might be used again with advantage, mixing three parts of wheat with one part of beans. Bran is occasionally used as a laxative, but it should be sparingly given. Beans are too nitrogenous to be used alone, and should always be mixed with more starchy food. Whatever grain horses are receiving, it should always be crushed or broken, but never ground into flour.

A certain amount of cut chaff should always be used with the corn, but this chaff should be from a half to three-quarters of an inch in length, so as to prevent any possibility of the horse bolting its food. The next point is, should it contain any straw? Personally I do not approve of straw chaff for horses, but I admit that a little sweet oat straw mixed with the hay may not do any harm. Wheat straw chaff should on no account be used. Straw chaff is used entirely for horses in some districts, and in others one seldom sees it at all, but I have noticed repeatedly that the cases of colic rise in number as the use of straw increases. For 9 years, with an average of 12 horses in the stable, I never had a *single case* of colic while I used hay chaff only or hay and a little oat straw, but on moving to another district, in deference to local custom, I allowed finely-cut wheat straw to be used, and had as a consequence three or four cases that winter. The colic ceased with the use of the straw, and I have seen this in other cases than my own. About 2 or 3 lb. of chaff is quite sufficient, and the rest of the hay should be given long.

The following are examples of rations for farm horses in full work :—

	lb.		lb.		lb.
Maize .	. 9	Oats .	. 7	Oats .	. 7
Beans .	. 5	Maize .	. 5	Wheat	. 6
Chaff .	. 2	Beans .	. 3	Beans .	. 2
Hay .	. 12	Chaff .	. 2	Chaff .	. 2
		Hay .	. 12	Hay .	. 12

The hay may of course be increased, half a pound of bran added, about 4 lb. of roots may be given in the spring, and if the work is light the corn can be decreased by 4 or 5 lb.

In districts where a stop is made at midday (and I believe this is the best plan), the horses should be watered immediately on entering the stable in the morning, and then receive one-third of their allowance of corn and chaff in two or three helpings or baits. On returning at midday water first and then give rather less than one-third of the corn and chaff, while on returning in the evening they should be watered again, receive the remainder of their corn, and be finally racked up with the long hay.

Where no midday stop is made, start by watering and giving about one-third of the corn and chaff as before, and put a small quantity into nosebags, to be taken into the field and eaten while the men are at lunch. On returning from work the horses are apt to drink too greedily, so give them about half an allowance of water, and then about 4 lb. of long hay to eat while the men are at dinner. When the men return, the hay will all have been eaten and the horses cooled down, so they may have as much water as they like, and then receive the remainder of their corn and chaff in several baits. Just before the horses are left, the rest of the hay should be given.

I have entered into these particulars because much discomfort, and often harm, is caused to horses by keeping them without water, and by giving large drinks of water on the top of newly swallowed food. Regular meals are also essential to health, and the shorter the interval between them the better. I do not like sloppy food for horses, but I believe it is a good plan slightly to damp the chaff just before using it, as this prevents dust and ensures the corn and chaff being eaten together. The hay should be of good quality, free from mould or dust, and must not be strongly heated.

I prefer a clover or clover and rye-grass hay myself, but of whatever kind it may be, it should have been cut before it became old and benty. Grooms have two peculiar notions about hay for horses: first they like it hard and benty, and second, they like it not less than a year old. Now usually one is able to detect some glimmer of the truth, though not always the whole truth, in these old practices, but in this case I have so far been able to see no reason. It has been proved beyond doubt by analysis, by carefully conducted experiments, and by experience, that the flower-stalks of grasses, *i.e.* the bents of the hay, if allowed to stand after flowering, are very little better than straw and are exceedingly indigestible. It is the young and leafy portion of the hay which contains most of the nutrients and a large proportion of albuminoids. It has also been shown by experiments that the digestibility of hay *decreases* with age, and as both points make the hay less digestible, it is difficult to see where is the benefit. I am inclined to think the practice is due to a survival of the notion that hard food was necessary for hard work, but as we know now that easy digestibility is essential to hard work, these practices should be relegated to the same category as pegging shrew-mice in trees for the cure of certain diseases. I am perfectly certain that attention to these details would often result in increased health and efficiency in many stables.

Horses should always have at least an hour and a half's feeding before going out to work in the morning, and a point often neglected is that every trace of the evening's meal should be removed from the mangers before putting in the fresh feed. Horses properly fed should have nothing left but a few hard or distasteful scraps, and these should be scrupulously removed. Every horse should be thoroughly well curry-combed and brushed at least twice a day. Horses at work in wet weather, especially on heavy land, get their legs plastered with mud. In the case of nags which are clipped and horses having practically no hair on their legs, this mud will soon dry, and can be brushed off; but with hairy-legged horses this is impossible, for much of it will neither scrape nor brush out of the hair, even after standing all night. What shall be done ? Shall it be *washed* off or left on ? I have tried both plans: at one stable the horses were always walked through a shallow pond on their way home from work, and in the other they were not. I never perceived any difference, except that the washed horses were the cleaner-looking. Veterinary surgeons are much against washing horses' legs, and, I think, rightly so in the case of clipped horses, for the rapid evaporation and consequent chill to the skin must be very great, while the conditions are altogether unnatural; but in the case of hairy horses, used to getting their legs soaked with dew, the skin is kept warm by the coating of hair, and the evaporation is much more gradual.

The building in which our horses are kept matters little, provided it is roomy, light, well ventilated but free from draughts, and scrupulously clean. To keep a stable thoroughly clean the first essential is a good sound floor, which should be of some non-absorbent material if possible. The strong smell of ammonia in many stables is due to the action of a bacterium which lives in the cracks and dirt of the floor, and converts the urea of the urine into volatile ammonium carbonate. Its action can be prevented by thorough cleanliness, and, in the case of absorbent floors, by sprinkling the floor before littering down with a handful or two of gypsum or superphosphate.

The hints which I have given on general management and feeding will be found, I think, as applicable to horses worth 500 guineas as to those worth 50. It is one of the greatest possible mistakes to suppose that, because horses have a pedigree, they must therefore be fed differently from, and pampered more than, ordinary work horses; but it is very often done, greatly to their detriment. We may certainly give them more zealous care and attention, but the requirements for health and strength are the same in both cases—a proper allowance of good food and plenty of exercise. Numbers of valuable horses are ruined every year by over-feeding; barrenness, fevered feet, skin eruptions, and other complaints being the result.

The rations I gave for working horses allow sufficient corn for most purposes, but for large breeding mares they may be increased by 2, 4, or 6 lb., care being taken to preserve the proportions. When horses are at grass or lying idle much less must be given.

The same rule applies to stallions, and when receiving a full allowance of corn, they must be vigorously exercised. A few minutes is not sufficient; it should be a walk of at least four miles a day. Fat stallions in flabby condition are useless as stock-getters, and as common as flies, but that which their work requires is good, firm, healthy condition. For breeding purposes there is nothing like a stallion that works in the team, but unfortunately fat catches the eye of most people and is therefore popular.

Stallions should be kept in large, light, and airy loose boxes, and if this can open into a small open yard so much the better. It is a bad plan, I am sure, to keep stallions shut up in nearly dark boxes, for it is my experience that the most tractable and best-tempered horses are those that are so placed as to be able to see all that is going on around them. They do not then become excited in the presence of strange men or horses. Properly used stallions may be turned out in well-fenced paddocks after the season much oftener than they are, greatly to the benefit of their feet and legs.

In spite of the fact that fat horses are quite useless for hard work, it pays the farmer to fatten the horses he has to sell. Fat often covers a multitude of faults, but while buyers are willing to give £30 or £40 for a layer of fat an inch and a half thick over a horse's ribs, so long will it pay the farmer to fatten what he sells. We must beware, however, of trying to fatten a horse too rapidly, unless we are prepared to risk fever and other ills. It is very common to give horses large quantities of linseed cake, peas, and beans; but such foods are too highly nitrogenous, and should be balanced with maize, wheat, or barley. A slightly larger allowance of the ordinary working mixture of corn is all that is really required; and if the horse is well exercised and trained at the same time, no harm will result.

I have known some farmers make a very good thing of buying good colts and selling them for town work after a year or two in their teams, but this requires very special knowledge and experience. I am of opinion that a certain amount of judicious breeding is in the long run the more profitable, and is undoubtedly the most interesting.

CHAPTER X

Flocks and Herds

IN selecting cattle for the farm we cannot complain of any paucity in the number and variety of the breeds. Those recognised in the prize-list of the Smithfield Club as beef-producing cattle are Devons, Herefords, Shorthorns, Sussex, Red Polled, Aberdeen-Angus, Galloways, Welsh, Highland; and, as small cattle, the Kerry, Dexter-Kerry, and Shetland.

The breeds generally recognised as dairy cattle are Shorthorns, Red Polled, Ayrshire, Jersey, Guernsey, South Devon, Kerry, and Dexter-Kerry.

The Longhorn is of historical interest as the breed upon which the famous Blakewell of Dishley gave such evidence of his skill as a breeder, and though the breed had nearly died out, an effort is now being made to revive it. Several farmers in Essex also possess herds of the black and white Dutch cattle, which they are endeavouring to keep pure.

All these breeds have their own special points of excellence for particular places and purposes. I remember reading a book on live-stock, in which each breed was described by some well-known breeder of that particular kind; and the result, with one or two exceptions, was that each breed was in turn proved to be the best. Now if one walks through the Royal Agricultural Society's showyard, and carefully compares them, one will learn more about the appearance and points of the different breeds than by any amount of description. The Dairy Show, with a record of the milking trials, will firmly establish the claims of the best milking breeds; and the Smithfield Show, if one carefully compares the weights of the cattle at the different ages, will give one much information as to the ability of the breeds quickly to produce beef for the butcher.

In practice the farmer has to be influenced by the purpose for which he requires the cattle; and if he is buying for beef production, he buys those breeds or crosses which he knows fatten rapidly and sell readily, and if it is milk he requires, he ignores breed almost entirely, and buys those individual cows which look like milkers. If, however, he wishes to keep a few cows for dairy purposes, to breed from them, and eventually fatten their progeny for the butcher, he must be very careful to choose a kind capable of producing both milk and beef.

The cows of some of the prime beef breeds can only produce enough milk to rear their own calf, and it then becomes important to consider if it will pay to keep a cow for a year and take the risk of depreciation and loss, for the sake of one calf. There are only two cases in which I think this is possible : where a considerable run of very low-rented grass is available, and where there is the probability of selling some of the progeny for breeding purposes at a much higher price than butcher's value. This point must be very carefully considered; for I am of opinion that if accurate accounts were kept, many breeders, and particularly gentlemen farmers, would find that cattle reared in this way resulted in a considerable loss.

The great popularity of the Shorthorn, the growing taste for the Red Polled, and to a less extent of the Devons, Welsh, and the two Kerries, is due to the fact that by them you can get a beef-producing calf, and several pounds' worth of milk into the bargain. These dual-purpose cattle are certainly the most profitable for the ordinary farmer under the conditions prevailing in this country.

The production of milk is often a very profitable part of farming, but the choice of breeds may be regulated to some extent by the particular dairy product we wish to turn out. In milk-selling we are anxious to obtain the largest possible yield of milk conforming with the government standard of not less than 3 per cent. of butter fat and 8.5 per cent. of other solids. All the dual-purpose breeds I have mentioned are suitable for this purpose, and we can add to them the Ayrshire and the Guernsey. A good cow of any of these breeds should be capable of producing from 600 to 1200 gallons in a season's milking when properly fed, and I do not think a cow is worth keeping as a milker whose yield is less than 500 gallons. Not long ago I possessed a shorthorn cow that gave 230 gallons, or a little over a ton of milk, in six weeks, although she had no other food than grass, and I know that her calf grew into as fine a beef beast as could be desired. These milking breeds are equally good for such an industry as cheese-making, where a large quantity of milk of fairly good quality is required. For butter-making the case is somewhat different, for although a large yield of milk is desirable, the quantity of butter fat it contains will determine the amount of butter produced. A cow giving 30 gallons a week of milk containing $2\frac{3}{4}$ per cent. of fat would yield hardly $8\frac{1}{4}$ lb. of butter fat, while a cow giving 20 gallons containing $4\frac{1}{2}$ per cent. of fat would yield 9 lb. of butter fat per week. The Channel Island cattle are undoubtedly the best butter-producers, Jersey cows sometimes giving milk containing as much as 6 per cent. of butter fat. The great drawback to these breeds as farmer's cattle is the fact that the bull calves, and old or barren cows, are practically valueless.

It certainly does not pay a farmer to keep inferior cows of any kind, and he should periodically weigh the day's milk given by every cow to ascertain the yields, and he should at the same time also test the percentage of butter fat in the morning and evening milkings. This is very easily done by one of the modern milk-testers; and all cows giving an inferior yield, either in quantity or quality, should be got rid of at the earliest opportunity. An inferior cow eats very nearly as much, and costs exactly the same for labour as a good one, and no amount of feeding will ever turn a bad cow into a good one. We can, by correct and generous feeding, make a cow give more milk up to a certain limit, but I cannot too strongly emphasise the

fact *that no amount of feeding will ever materially or
permanently alter the quality of the milk.* This has
been abundantly demonstrated by numerous experiments
in nearly every country in Europe and in America, those
at Copenhagen alone having involved over 1600 cows.
It has been found that any sudden alteration of diet
produces at the time a slight change in the quality
of the milk, but that, in a few days at most, the milk
returns to its normal composition with a change in
quantity. If farmers tested their milk as they ought,
they would soon be convinced of the truth of these
statements.

Milk is formed by the actual breaking down of the
living cells lining the milk glands, the broken-down cells
and their contents flowing into the ducts as milk, so
that it is difficult to perceive how the food can make any
difference in the composition of these living cells. It
is easy to see that a cow in bad health may give milk
which is not normal in its composition, and we know
that certain smells and flavours may permeate the blood,
and thus gain admittance to the milk. It is also known
that such a food as cotton cake gives rise to a butter of
harder consistency than that produced by linseed cake,
but the reason of this has not yet been ascertained.
It has been clearly shown, however, that although you
cannot alter the percentage of butter fat from that
normal to any particular cow, yet we can make a cow
give more butter by increasing the quantity of milk.

Milking cows require a good deal of care in their
feeding if the greatest possible return is to be obtained.
The cow has to maintain herself, often to nourish a
growing calf, and at the same time to give milk, so it
is evident that she requires a good deal of food, and

that this food must contain a large proportion of albuminoid material, as well as carbohydrate and fat.

Good sound hay and oat straw should form the foundation of a ration for cows, indeed for most cattle. The hay should be cut into chaff an inch long, with about an equal weight of the oat straw; but if clover, sainfoin, or lucerne hay is used, rather more straw may be added. From 10 to 14 lb. of this chaff per head per day is mixed with the roots or other food, and from 7 to 10 lb. of either hay or straw should be given uncut.

Swedes and mangolds may be given in fairly large quantities—say from 20 to 60 lb. per day; but common turnips and cabbages must be fed more sparingly. If roots are plentiful, 50 lb. a day is not too much, but it must be remembered that roots used with straw do not supply nearly enough albuminoid material for the requirements of any animal. Decorticated cotton cake or meal is the most highly albuminous food we can use, and it is often the cheapest; so taking this cake as a standard, we may say that 5 lb. will make up the deficiency in the roots and chaff. This simple ration will then read as follows: Chaff, 14 lb.; swedes or mangolds, 50 lb.; decorticated cotton cake, 5 lb.; long hay or straw, 8 lb.

If we happen to be short of roots, we may use 1 lb. of wheat, barley, or maize meal for every 7 lb. of roots, giving, for example, 30 lb. of roots, 5 lb. of decorticated cotton cake, and 3 lb. of wheat meal. We may substitute silage for roots practically weight for weight, but silage varies much in quality. If brewer's grains are used, it is necessary to decrease the cotton cake by about 1 lb. for every 10 lb. of grains, thus,

20 lb. of roots, 20 lb. brewer's grains, 3 lb. decorticated cotton cake, 4 lb. of wheat or barley meal. This ration, with the hay and straw, would be an excellent one for heavy-milking cows where roots were rather scarce. When a considerable quantity of straw and roots forms part of the ration, there is no food we can successfully substitute for the decorticated cotton cake, owing to its high percentage of albuminoids; but, with a larger quantity of hay and less roots, we can use beans to supply the albuminoids, and linseed to supply the oil — thus, 24 lb. hay, 20 lb. roots, 5 lb. beans, and 1 lb. linseed.

The changes which can be rung on our common foods are very many, but these will, I think, give an idea of the way they may be manipulated. No milking cow or fattening animal should ever be fed with an exactly weighed allowance. The proportions should remain constant, but heavy milkers, at any rate, should have as much as they will eat. Drying-off cows should get less, and the albuminoid food, beans or cake, should be decreased.

Unless you are going to sell a cow, never let her get fat, or you will at once be open to that dreadful scourge of the cow-keeper, milk fever. The great predisposing cause of milk fever is too rich food, too much cake, and consequently too high condition. I have been through dairy herds, numbering up to 700 cows, in various parts of the country, and what I have seen and heard only confirms my own experience. In twenty years, with numbers varying up to over fifty cows in milk, I have never had a single case. It may have been luck, but I never bought a fat cow, nor kept one that got fat.

Cows kept for breeding purposes only, must not be

fed in the way recommended for heavy milking. Such
cows should be kept simply as stores, and should drop
their calves in early spring, so that they and their calves
may be turned out to grass as soon as possible, and all
cake-feeding avoided. In the case of the non-milking breeds, the calf
is usually capable of dealing with all the milk the
cow can give, and the mother and calf are simply
left together. With cows which have a fair supply
of milk, but are not intended for the dairy, they may
be made to suckle another calf besides their own.
This is often an excellent plan, for many cows and
heifers of the milking breeds will bring up the two
calves practically as well as one, and the increased
return helps to pay the keep of the cow. Strange
calves are only detected by the cow by their smell,
and so if we keep her own calf and the stranger shut
up together in a little pen, letting them out three
times a day to suck, in a few days they will both smell
alike, and may then run together with the cow. There
is no reason why many a pedigree cow should not rear
another calf in this way, for the slight difference it would
make to her own calf would be more than paid for by the
increased value of the other, while there is no doubt
that the cow at grass can regulate her milk-supply to
some extent by the demand.

In the case of cows intended for the dairy, the calves
must be taken away and reared by hand. It is certainly
much better for the calf to be allowed to take all he
requires from his mother for three or four days, but, in
any case, the calf must have his own mother's milk for
a day or two. Calves are very easily taught to drink
out of a pail, and should have at least 4 quarts of new

P

milk, given in two or three meals a day, till they are six weeks old. At this age we can begin to replace the new milk by some other food if we wish, but I must strongly emphasise the point, that the treatment of the first three months determines the animal's character through life. The well-nourished calf rapidly produces flesh, covering the shoulders, back, and rump, with muscles which give that rounded, beefy appearance so desirable in animals intended for the butcher. This calf-flesh once lost can never again be recovered, and the starved calf, however it may be fed afterwards, will always be hollow behind the shoulders, and generally deficient in that firm fulness indicating the well-marbled flesh so sought after by the butcher. It is easy to see from this the great importance of good feeding for calves intended for beef; but I have already pointed out that beefiness is not a desirable characteristic of animals intended for milking, and that fatness tends to reduce the breeding powers of all stock. In the case of the female calves of the milking breeds, especially of the Channel Island cattle, it is not desirable to feed them too highly, but rather to keep them simply in growing condition. Attention to these points will lead to success in the market or showyard, and breeders of Jerseys will not have to complain of their heifers growing too large and coarse.

As calf-rearing on new milk alone is expensive, it becomes important to know how it can be replaced, and there is no doubt that substitutes can be found, which, if used with care, will give good results. After the age of six weeks the new milk can be gradually replaced by skim milk, but it is necessary to supply the fat removed, if good results are to be obtained. Cod-liver oil, cotton-seed oil and others have been recommended, but I think that gruel made from crushed linseed is preferable. I have, however, in rearing several hundred calves, tried nearly everything, and I have come to the conclusion that, on the score of economy and efficiency, there is nothing to beat Bibby's calf-meals—the Cream Equivalent for mixing with skim milk, and the Milk Equivalent for replacing new milk. It must be remembered that simply increasing the supply of skim milk will not nourish a calf, for it is the fat that it misses, and to give more milk without the fat is only to overload the calf's stomach with food that it does not require. Personally I have obtained the best results from using new milk, and gradually replacing the milk by gruel made from the calf-meal, so that at the age of ten or twelve weeks the calf is getting gruel only.

Calves will begin to pick a little sweet, soft hay by the time they are a month old, and shortly afterwards they should be tempted to eat a few crushed oats. It is a mistake to supply a food so rich in albuminoids as linseed cake to calves receiving skim milk; much better results will be obtained from oats or a mixture of one part of linseed to three parts of barley or wheat meal. For calves going off milk altogether a mixture of two parts of linseed cake to one part of barley meal may be given; or, a mixture of two parts of barley meal, one part of bean meal, and one part of crushed linseed. A calf should always have a rack of good hay to pull at as it likes, and at three months old it should be receiving about 1 lb. per day of one of the mixtures recommended, with about 1 lb. of slightly damped hay chaff. The quantity may be gradually increased, till at six months it is getting nearly double. After this the allowance of concentrated foods need only increase very slowly, as the calf can now eat roots and a considerable quantity of hay.

The American authorities on cattle-feeding recommend a more highly nitrogenous diet for young cattle intended for beef than for those intended for milking, and, with the allowances given above, this difference could be met, to some extent, by giving meadow hay to the dairy breeds and clover hay to the beef breeds.

I am fully convinced that the rearing of calves, when properly carried out, is a very profitable branch of stock-keeping, and I am also sure, as in many other branches, bad luck is only another name for bad management. The calf-houses should be light and airy, kept scrupulously clean, and lime-washed frequently. All pails used should be scalded, and the mangers and hay-racks cleaned out every day. Scouring and ringworm seldom appear among well-managed calves, and there is no more interesting sight on the farm than the groups of glossy-skinned, bright-eyed calves. They are so intelligent and inquisitive that one always regrets them growing into creatures that think about nothing but eating, just as one always regrets that kittens grow into cats.

Although calves that are going to be turned out to grass with their mothers are best born in the spring, autumn calves are the best for hand-rearing, for they will be six or seven months old, and ready to turn out as soon as the grass comes. By the following September they will be in splendid condition, and will winter well on straw and roots with a little hay and meal.

Spring- and summer-reared calves are much better when not turned out at all till after their first winter, for it is my experience that calves turned out after

June never do well even if allowed some meal. All the calves should be brought into their winter quarters by the end of September at latest, including those running with their dams. To leave them out later is only to risk "husk" and other ills.

The calves which are now weaned from their dams must be kept in good condition, for, if allowed to sink, it will cost far more to get them up again than it will to keep them fresh. About 2 lb. a day of one of the mixtures recommended for calves going off milk will be very suitable, and it should be given with 3 or 4 lb. of hay chaff, and a small quantity of roots, which may be increased to 15 or 20 lb. by the end of the winter.

The autumn-reared calves, which are now a year old, may be given chaffed oat or barley straw and roots, a little hay, and 2 or 3 lb. a day of a mixture of equal parts of bean and barley meal. It would be very difficult to attempt to prescribe exact quantities, and, as a matter of fact, the farm fodders should be given practically *ad libitum*, only the concentrated foods being given in measured quantities. Some discretion must be used as to these quantities, and I have tried to indicate that cakes alone are not always the most economical foods, although they can sometimes be bought of such composition and at such prices that they become excellent supplements to our home produce. One often sees extravagant quantities of concentrated food given without commensurate results, because the proportions are not properly adjusted.

The calves, being now from twelve to eighteen months old, it is necessary to make considerable differences in their treatment, according to the purpose for which they are raised. The heifers which are intended for breeding should be in good healthy store condition only, and may be turned out to grass in their second spring, to grow till big enough and old enough for breeding. Jerseys are often allowed to bring calves at two years old, but to obtain fine, healthy, well-developed cows, three years old is quite soon enough for calving. The increased value of such cows will, I think, more than pay for the loss of time.

The young bulls, as well as the steers and heifers intended for the butcher, must be differently treated. The bulls may be fed on a diet somewhat similar to that which the other yearlings have been receiving, i.e. 4 or 5 lb. of hay or straw chaff, with from 15 to 20 lb. of roots, and 2 or 3 lb. of concentrated foods, but they may have as much long hay as they will eat. Many growing bulls are given much more concentrated food than they can possibly require or make use of. A mixture of beans and oats in equal parts would be very suitable, or equal parts of linseed cake and barley. Beginning at 3 lb. a day, the quantities could be increased as the bull grew older; but avoid over-feeding—it is so detrimental to the animal's sexual functions.

Young bulls are best in open yards which have a shelter hovel, but at eighteen months old they should be either tied up or put in loose boxes. Rings should be put in their noses at twelve months, as it gives a means of control when required. The temper of bulls is always uncertain, and they should never be played with and petted or teased, or they may lose their inborn respect of man. An old stockman, very well known in the showyards in his day, and who had had hundreds of bulls through his hands, once told me that it was his

rule never to go in to a bull without a short whip in his
hand, but that it must never be used unless urgently
required, and he had found that bulls had more respect
for the clang of a whip than anything else. I have
repeatedly found this to be the case; but every animal
most respects the man who is not afraid of him. Bulls
used quietly and properly are never bad tempered,
though they may be unruly at times.

In the fattening of cattle, a very large number—I
might almost say the majority—of farmers have not yet
altered their methods to suit the new conditions of the
fat-stock markets. Until a comparatively recent period
it was not considered possible to begin fattening cattle
until they were three years old at least; indeed for
fattening on the rich pastures cattle of that age do
better and are still preferred. For stall fattening, however, things have entirely changed, and there is now
no doubt that fattening at an early age is the most
profitable method.

A visit to the Smithfield Club's Show will demonstrate
that there are several breeds, which can produce very
nice butcher's beasts at under two years old. The
animals at the show are of course picked specimens,
but the ordinary fat-stock markets will show that such
cattle realise prices equal to about £1 for every
month of their age. How many three-year-olds could
show an increase at anything like the same rate? The
weigh-bridge shows the feeder many important points,
among them the fact that the older the animal gets the
greater the amount of food it takes to produce each
pound of increase. Well-fed cattle gain in weight at a
fairly uniform rate of about 2 lb. per day from birth
till they are two years old, but after this the gain per

day from birth tends to get less, and the cost per pound
of the increase is double in the second year, and treble
in the third.

It stands to reason, then, that the earlier we can get
our cattle off the more chance we shall have of a profit,
and it is only by realising that such changes in pro-
cedure are necessary that we are able to keep pace
with the times. Some fifteen or sixteen years ago, I
remember visiting a beautiful and old-fashioned corner
of the Midlands, then eight miles from the nearest
station, where a charmingly old-world farmer of nearly
eighty showed me a stall of beasts, which, failing to
realise the price he expected at one Christmas market,
were being kept for the next. How delightfully re-
miniscent was this of the easy-going methods of the old
times; and as I silently reflected what the loss of this
proceeding must have been, I could not help agreeing
with the old fellow that times had strangely altered.
The alteration, with regard to live-stock at any rate,
is not all for the worse, for the newer methods enable
us to keep much more stock, to turn our money over
more quickly, and have, I believe, only changed the
profit from the corn over to the stock.

There seems to be no doubt whatever as to the
advantages of early and rapid fattening, and having
brought our calves up to fifteen or eighteen months old
in good beefy condition, the final fattening is not very
difficult. The rations indicated for dairy cows will be
found almost exactly in the proportions suitable to the
requirements of our young fattening cattle, but of course
they will not require more than from three-fourths to
five-sixths of the quantity. Dried brewer's grains are
much relished by stock, and might replace the wet

grains in the cow's mixture: 1 lb. of the dry should be used in place of 4 lb. of the wet grains. For the last six or eight weeks of the fattening it is an excellent plan to replace 2 lb. of the decorticated cotton cake, or 3 lb. of the beans, by 3 lb. of linseed cake, as the change is liked by the cattle, and linseed cake has a subtle way of adding fineness of touch to the skin. How long the final stage will take must depend on the previous condition of the animals and the care and skill with which they are fed. There is, however, little object in having them over-fat, for, except at Christmas, the butcher does not care for extra fat meat, and the last few pounds of fat cost the feeder much more per pound than any other. These fattening cattle may be fed in yards, or, I think, more economically, in boxes or tied up in sheds.

So far I have only supposed our young cattle to have been running at grass in the summer and brought into the yards or sheds in winter to be fed or fattened on hay, straw, roots, and meals. We can, however, fatten our cattle during summer quite as successfully, and perhaps even more economically, by the method known as " soiling," *i.e.* by carting green food to the cattle in the yards and sheds. The method has great advantages, enabling us, by growing suitable green crops, to keep a far larger number of cattle than on pastures alone, and, in fact, to keep a considerable quantity, where no pastures are available. Cattle of all ages do exceedingly well upon such green food as tares, clover, lucerne, rye grass, maize, and cabbages, very much better than upon hay, straw, and roots, while the largeness of the crops enables a small area to support a comparatively large head of stock. The drawback generally urged is the labour necessary to cut and cart to the yards the rather large quantity of fodder required. One man with a horse and cart could cut and bring to the yards all the fodder necessary for a very large number of stock, but that a horse and man should be seeing to stock in the summer is one of those changes to which the farmer finds it hard to become reconciled. Full-grown cattle would require about 100 lb. of green stuff per head per day, so that, even if there were thirty head in the yards, the labour would not be very great to bring in and give some 25 cwts. of green food in two feeds. For fattening cattle a small allowance of concentrated food would be necessary—say 2 lb. of decorticated cotton cake and 2 lb. of barley meal; although, if the fodder were entirely from leguminous crops, 4 lb. of barley meal would probably meet the case.

Partial soiling might be practised far more often than it is in the case of a limited area of pasture, for the animals might be kept in the sheds during the hot part of the day and fed with, say, 50 lb. per head of green stuff, which would enable double the cattle to be carried. When we consider that 15 tons per acre or more can easily be grown on the ploughed land, we can see how a few acres judiciously planted could materially help out our pastures. The surplus fodder of any crop when another was ready could then be consumed by sheep on the land, made into hay, or put into the silo for winter use.

This, then, gives an idea of what may be called "intensive" stock farming: laying our plans to produce rapidly the greatest possible amount of beef from the land at our command.

Very large numbers of cattle are still fattened upon grass in the grazing districts; but this industry, although still profitable in the hands of the keen judge of stock, is not what it used to be. I have heard old graziers speak of the times when they could buy a couple of hundred good Irish stores at £8 or £9 a head, and in sixteen or eighteen weeks they would all have been sold fat at from £18 to £20. That, of course, was farming made easy, and it is no wonder that in such districts every possible acre was sown down to grass. Times have altered now, for although beef is quite as dear, stores are much more valuable. The buying in of stores to fatten is now a precarious business whether it is intended to fatten on grass or in yards.

Numerous experiments have shown that in the yard-fattening of store beasts it is exceedingly difficult to make any profit, whatever the food used, the determining factor usually being the relative buying and selling prices. The general conclusions with regard to foods have been, that with an allowance of hay and straw chaff such as was given for cows, swedes and mangolds may be used to any extent, for the more roots the cattle eat the better they seem to do. As concentrated foods, linseed cake alone gives good results, but in point of economy it is beaten by a mixture of decorticated cotton cake and maize or dried grains half and half. The concentrated foods should begin at 6 lb. per head per day, increasing to 8 lb., and afterwards to 10 lb. Undecorticated or common cotton cake gave poor results in every case ; it is indeed a food little to be recommended, except perhaps for feeding to adult cattle upon grass, and it should be carefully avoided for young stock of all kinds. I have heard of many deaths of calves and lambs attributable to its use.

Suitable mixtures of home-grown corn, such as beans,

barley or wheat, and perhaps a pound of linseed to supply oil, will give practically as good results as bought cake, and are often cheaper; but in every case it is difficult to be sure of a profit. As long as the four-course rotation of cropping holds, the fattening of bullocks in this way must take place; but if the farmers would modify their cropping so as to produce their own beasts, they would, I am sure, secure larger profits.

With sheep as with cattle there is no lack of variety of kinds, there being quite two dozen distinct breeds. In selecting any particular kind it is certainly best to choose one which is fashionable in the district, for such a breed has probably been found to thrive well, and there is a better chance of selling it profitably. The mountain breeds are of course not suitable for the kind of farming we have been considering, but any of the lowland breeds can be successfully kept on the ploughed land. Wool is now of so little value that it can be entirely disregarded, and the production of good and saleable mutton should be our chief aim. It is not necessary even to keep a pure breed for this purpose, but the farmer who keeps a pure-bred flock has perhaps a greater source of interest, and besides has the opportunity of selling at higher prices for breeding purposes.

If you have a pure-bred flock, try to keep it first-class, for there is very little profit, I am sure, in selling third-rate rams. The same amount of food given to ordinary sheep would often produce mutton of greater value than many of the rams I have seen sold. As in the case of horses and cattle, the pure-bred flock requires no different feeding or treatment from that of any other flock properly kept. All sheep pay best for careful attention and generous feeding, but this feeding should be from the produce of our soil, not from the pocket.

The farmer who simply buys in sheep to consume any particular crops he may happen to have grown, begins his sheep-keeping at the wrong end, and unless he makes a considerable allowance for the value of the manure, it is doubtful if he often makes a profit. To breed the sheep and to grow crops on purpose for their use is certainly the better plan, for then whatever increase our sheep make is a gross profit to the farm, and we have the manure into the bargain.

It is possible to breed some sheep on any kind of land, but, as I have already pointed out, the lighter soils are the best suited for large breeding flocks. There is, in my opinion, nothing more interesting than the planning of the crops on such a farm so that a succession of suitable foods may be provided all the year round, on which the sheep may be kept in flourishing condition, with practically no buying of any kind.

The time at which we expect our lambs makes a good deal of difference to the provision we require to make, for, omitting the Dorset breed which may bring their lambs in September, the usual times of lambing are January, February, and March. Where the sheep are intended for showing, or the production of fat lambs for the spring market is the object, then lambing cannot take place too early in January ; but we must remember that the earlier the lambing the more expensive will be the keeping of the sheep. Ewes lambing at a time when green food is scarce must be fed on hay, roots, and corn or cake ; and if the lambs are to be pushed on as rapidly as possible, both they and their mothers must be given

concentrated foods for a considerable period. It undoubtedly pays to feed all young animals well; but the
question arises, does it pay to produce lambs at a time
at which they must be entirely hand-fed, when by producing them six or eight weeks later they will quickly
be able to support themselves on the crops growing on
the farm? Personally I incline to the later period of
lambing—that is, from the end of February to the end of
March; for by that time in most seasons we can reckon
on having crops of thousand-headed kale ready for use,
to be followed by tares and rye, spring cabbages and
other food.

Lambing time is one of the most interesting periods
of the farmer's year, and if proper provision has been
made there need be little anxiety; for, as I have so
often said, good luck follows good management. For
the keeping of sheep an experienced shepherd is a
necessity, but there are a few points which concern
the master's management on which I must give a few
hints.

Ewes must never be kept entirely upon roots previous
to lambing, for roots alone do not provide the proper
nutriment necessary for the growing lambs, and the
deaths of more ewes at lambing time are attributable
to this cause than to all the others put together. All
pregnant animals not properly fed draw upon their own
bodies to supply as far as possible the requirements of
their growing young, and if this strain is long continued
excessive weakness ensues. In-lamb ewes are much
the best at grass with a few roots and a rack of clover
hay provided, but this is not always possible, and when
penned on roots care must be taken to provide as much
clover or sainfoin hay as they care to eat, as well as a

lump of rock-salt for licking. Do not expect ewes to thrive on the shells of roots left by other sheep, but if the ewes have cut roots and some leguminous hay to supply the necessary albuminoids, they should require no corn till within a fortnight or so of lambing.

A lambing-yard must be provided in a sheltered position, but under no circumstances should this yard occupy the same position in two consecutive years. This is of the greatest possible importance, for it is certain that the germs of various diseases can remain active during a year or more, and I have known several cases of bad luck in the rearing of lambs attributable to the use of permanent lambing-yards.

If there are many ewes to lamb, some one should be with them night and day ; and it is remarkable how the ewes seem to lose all fear of man at this period. All dogs should be fastened up; and although the ewes quickly detect a stranger, yet people they know may step over their recumbent forms and handle any of them without causing any disturbance. About the second day after lambing, if the weather is open, the ewes and lambs should be drafted out into a sheltered pasture and fed with roots, clover hay, and a pound of beans, or three-quarters of a pound of decorticated cotton cake per day. Lambs are very hardy, and at two days old can as a rule stand any kind of weather ; but weakly lambs should be protected from rain or sleet, which tries them much more than the hardest frost. After a fortnight on the pasture they may be penned on the thousand-heads which should be ready for them, and the dry food continued as before.

In the eastern and southern parts of England at any rate, it is a common custom to allow swedes to remain

in the ground and sprout in the spring to provide early
green food. The practice is, I am sure, a bad one.
The sprouting roots provide a comparatively small
quantity of food; they are not greatly relished by the
sheep; and all the nutriment comes out of the bulb, so
that the swede itself becomes useless as food. The
principal objection seems to me to be this: Why go to
the expense of growing so troublesome and uncertain a
crop as swedes, when a far greater weight of thousand-
heads can be grown at much less expense? The
expense in providing food for our stock makes the
greatest possible difference in the ultimate profits, and
although we must give sheep some food rich in albuminoids while they are eating roots or kale, much of
this expense can be saved too when we get them on to
a leguminous crop. When feeding upon tares, clover,
trefoil, or sainfoin, the ewes should require no cake or
corn at all; and for the lambs, a mixture of beans and
barley or wheat, half and half, or decorticated cotton
cake one-quarter to three-quarters of barley or wheat,
should provide all that is required. When kept in pens
the lambs should be allowed to run through hurdles
with upright slats, and they will soon learn to eat their
allowance of corn apart from the ewes. The quantity
of corn given to lambs will depend on how they are
doing and the purpose for which they are required, but
it need not in any case exceed a quarter of a pound per
head before weaning.

Lambs should be weaned at from sixteen to twenty
weeks old, there being no gain in allowing them to
remain with the ewes too long. As soon as weaning
has taken place the ewes may be put on short commons
for a few days to dry up the milk, and can afterwards

be used as the scavengers of the farm to clean up the remains of crops which have been used for other purposes. The lambs may be kept growing well and in good condition on the aftermaths of sainfoin or clover, on spring-sown tares, cabbages, and other crops; and we must bear in mind that sheep must not be kept too long on any one kind of food, but do better with repeated changes. This short description gives the general idea of the way in which well-grown sheep can be produced, sheep which should be in such condition that they will require a very short period of special fattening; but there are a few other incidental points to be considered. Lambs should be castrated and have their tails cut at as early an age as possible: they are not too young at a fortnight old; but where some of the male lambs may be required as rams, the tails should be cut and the castration deferred till one is better able to judge of their points. The ewes with twins may be separated from those with single lambs, so that both they and their lambs may receive a slightly larger allowance of corn.

Dipping is an important operation with sheep, and all ewes should be dipped once and the lambs at least twice during the summer. Not only does dipping destroy the vermin which cause the sheep so much annoyance, but it improves their appearance and prevents to a considerable extent that pest of the sheep farmer, the maggot. No dip will entirely prevent the green blow-fly, which is the principal offender, from depositing its eggs in the wool, but an arsenical dip will often kill the maggots as soon as they hatch out, and in my experience the best preparation for killing the maggots when found in the wool is a weak solution of "Jeyes'

Q

Purifier." Foot-rot often gives a great deal of trouble, but it is largely preventible. It is caused by a bacterium which lives upon the softer parts of the horny covering of the foot, and, if left unchecked, it eventually gets inside the hoof and completely destroys the connections between the horn and the tissues of the foot. Although the bacteria may be common in the soil, they can gain no admittance to a perfectly healthy foot; but under the conditions in which sheep are kept on farms the hoof grows much faster than it can be worn away, and inturned and ragged edges of horn appear. These ragged edges retain dirt and moisture, and provide just such a lodgment as the bacteria require. It will be found that if the sheep's feet are carefully pared two or three times a year foot-rot will largely disappear. Copper sulphate, either as a strong solution or as an ointment, is the best cure for individual cases.

Ewes as a rule should not be kept after their fourth lambing, for they soon begin to lose their teeth. Directly after weaning the lambs, all ewes which have reached this age and any others we may wish to get rid of should be drafted out of the ewe flock and run with the lambs, so that they may get fat for selling. To keep up the flock, a sufficient number of the best ewe lambs must be selected each year and kept round for breeding. Sheep will breed in their first year, but it is usual not to allow them to drop lambs till they are two years old. In these days of early maturity, however, there is no reason why a certain number of early, well-grown lambs should not be allowed to breed, so that they will bring their young at about fourteen months old. I have never been able to detect any harm resulting from the practice when the sheep are kept well.

About a fortnight before the ram is put with them the ewes should be given better feed. A rather thin ewe just beginning to do well is more likely to breed than a ewe in good condition, but the tale that mustard or any other such feed will influence the breeding is, I think, a myth. One ram can manage fifty ewes, but it is my experience that such a number is too many for each ewe to receive proper attention. The farmer with twenty or thirty ewes always has a larger percentage of twins than his neighbour with 100 or 200 ewes. It would be difficult to say exactly why, and there is probably more than one reason; but to run a smaller number of ewes with each ram, and to change the rams if possible at three weeks, would, I think, effect an improvement.

Lambs which have been kept well should be ready for turning into mutton at nine or ten months old; in fact, kept in the way I have mentioned, some may be ready sooner. For fattening on a leguminous crop, a mixture of beans and barley in equal parts will be found excellent, beginning with half a pound per day and increasing to one pound. On roots, the best corn is undoubtedly the same quantities of decorticated cotton cake and barley. When feeding sheep of any kind on roots they require a small quantity of dry food, about half a pound per head per day of clover hay chaff being probably the best. The corn should be mixed with this, and the roots should always be cleaned and sliced. The extra labour will be more than paid for by the increase in mutton.

In preparing rams for show or sale, shepherds nearly always feed a diet much too high in albuminoids; such as linseed cake and peas. Not only is such a food bad for the health of the animals, but it has been proved that they fatten better on the mixture of cotton cake and barley recommended above, which is also much cheaper. Although sheep are, I believe, the most profitable class of stock we can keep, yet economy of production is as important as in any other case.

In this brief sketch of sheep-keeping I have tried to avoid those points upon which common practice is agreed, and to emphasise those points upon which practice might be improved.

Some of the largest and best flock-owners I know keep their sheep upon the general lines I have indicated, and with the best possible results. The number of sheep as well as other stock which some of them contrive to keep in proportion to their land is remarkable, and one cannot help feeling convinced that this style of stock-keeping is really far more profitable than permanent pastures. On the pastures a few sheep fit in well with cattle, as, owing to their different habits of grazing, they keep the grass fairly level between them ; but if there are many sheep they will certainly pick out all the best of the herbage. When on the pastures sheep should be changed from one field to another every few days, and they will do very well, but the ploughed land is the place for sheep during the greater part of the year.

The keeping of pigs is, no doubt, often a profitable part of farming, but the place of the pig as the farm scavenger has become largely a thing of the past, modern practice having completely altered our ideas of profitable pig-keeping. The method which is now rapidly disappearing was to breed a certain number of pigs, which were brought up as stores, and allowed to run about and get a living on any kind of oddments. These were the scavengers of the farm, and when they had reached a certain age and size, they were put up and fattened to produce large and heavy bacon pigs. Such large pigs are not now required, and realise much less per pound than smaller sizes; besides this, every feeder who keeps a strict account of the foods he uses, knows that it requires more meal to produce a pound of meat in a large pig than in a small one.

The pig which pays best to produce is one having a carcase-weight of from 90 to 100 lb., and in producing such pigs we must take care that they are never allowed to become stores; they must be kept fat from weaning to killing. These pigs are splendid utilisers of the waste products of the dairy, although personally I agree with Fleischmann in preferring the calf as a utiliser of skim milk.

Breeding sows and young pigs intended for breeding are much better when allowed a good deal of liberty, and if they are rung to prevent rooting, they will get most of their living in a grass field. I have several times been on one large pig-breeder's farm where thirty or forty sows may be seen any time during the summer industriously grazing in the meadows. They are allowed about a quart of beans per head per day, thrown down on a bare patch of ground. When unground corn is given to pigs it should always be scattered on the ground, so that it cannot possibly be bolted in mouthfuls. Pigs of this class may be kept in the yards and fed on tares, clover, cabbages, and other green food during summer, and on mangolds during winter, but they will require some corn as well. As these pigs are either growing or carrying young, they require a food fairly rich in albuminoids, and beans are about the best form in which this can be given. Maize, being very poor in both albuminoids and ash, is the worst food that can possibly be used for growing pigs, unless mixed with some other food which will supply its deficiencies. The ordinary swill upon which the "old sow" is often condemned to be kept is usually very poor stuff; for the enormous quantity of liquid she has to consume, all has to be warmed up to the temperature of the body, and for this a great deal of the food must be used to supply the necessary heat.

Sows should never be fat at farrowing time, but they should be in good condition, as they lose flesh very rapidly when suckling, even with the best of food. After the first two or three days from farrowing the sow should be liberally fed with a considerable variety of foods. Meal mixed with slightly warm water is the best food, but it should not be too sloppy, and water for drinking should be provided separately. At about three weeks old the little pigs should be allowed to run into a pen away from the sow, and be tempted to help themselves to a little skim milk or thin gruel of milk and meal.

The best meal for suckling sows and young fattening pigs is a mixture of wheat, barley, and perhaps oats, with about one-sixth part of beans or peas. Middlings or pollard may be used if cheaper than wheat; but' bran is an unsatisfactory food for pigs, being rather indigestible.

All pigs kept in pens require some exercise, and if they cannot be let out for a short time, the best way to provide exercise is to put a barrowful of sods in the stye, and throw a handful or two of corn amongst them now and then. Not only will the rooting for this corn provide healthy exercise, but it will also supply the pigs with grit to eat, a matter very necessary for health. Attention to such points as these, and the provision of sufficient albuminoid in the food, will prevent any fear of the sows eating their young, and will secure healthy, well-grown litters.

At eight or nine weeks old the little pigs will be feeding well, and as the sow will be giving very little milk, they may be weaned. Now is the time when it pays to keep the little pigs going on in good fresh condition, and, if it can be managed, no food suits them so well as barley or wheat meal mixed to a cream with skim milk. Pigs cannot produce lean meat and bone economically on such grain as wheat, barley, or maize alone, and if skim milk cannot be given to supply flesh-formers, then beans or peas must be added.

It will be found that newly-weaned pigs will require about $2\frac{1}{2}$ lb. of meal per head per day, and that this quantity will increase, till at the time they weigh from 100 lb. to 150 lb. they will require 5 lb. per head. As a gallon of skim milk contains about 14 oz. of nutritive matter, less meal will of course be required when milk is given. When properly fed, these young pigs will put on a pound of increase for every $3\frac{1}{2}$ lb. of food eaten, but as they get older they want more—pigs of 100 lb. weight requiring $4\frac{1}{2}$ lb., and those of 300 lb. taking $5\frac{1}{2}$ lb. for every pound of increase.

Although a few store pigs will practically get their own living on a farm, and will fatten rapidly when well fed, it is evident that where many pigs are kept it pays much the best to produce pork at as early an age as possible; indeed this plan is adopted by all the largest and best pig-keepers.

There is no animal kept on the farm so generally mismanaged as the pig. Because he can manage to get along under the most adverse conditions he is allowed to make shift on a treatment which would mean illness or death in almost any other animal. The store pig has to drink gallons of ice-cold wash, and very often vainly seeks a little straw in which to warm his chilled frame. The fattening pig is given a pudding-like mess of meal, so dry that it must suffer agonies from thirst, and half its food passes through it undigested. Growing pigs are fed on only one kind of meal, often so deficient in ash and flesh-formers that stunted growth, rickets, and fevers are the result. That much-dreaded disease, "swine fever," is, I am convinced, greatly encouraged by the unhealthy feeding and filthy surroundings. It is a very infectious disease, but, as with all other animals, those in an unhealthy condition are most liable to infection.

The pig may not be so interesting, intelligent, and cleanly as the other farm animals, but he is given no opportunity of developing these qualities, and he certainly provides us with a means of converting our low-priced corn into more profitable meat, and thus becomes useful.

Poultry are usually considered to deserve a place among the live-stock of the farm. I am, however, exceedingly sceptical as to the profit in the majority of cases, although the want of profit may very probably be put down to lack of knowledge and skill.

I have seen poultry kept very successfully in a

yard divided from another person's field by a rather gappy fence, and not long ago a lady was showing me some very fine ducks which she was selling at 7s. a couple at eight or nine weeks old. The poultry were this lady's perquisite and resulted in a considerable profit, but the husband told me a few minutes afterwards, with a merry twinkle in his eye, that he supposed it must be the rats, for his meal disappeared whether he had any pigs in the stye or not. Unfortunately for my belief in poultry-keeping as a profitable adjunct to farming, I have always found some special circumstance which contributed to the profit. Practically all the villages where poultry-keeping has become an industry are possessed of commons or large greens on which the poultry can be run. Every villager possessed of a garden can keep a few fowls at a profit on the scraps of his house and garden ; and if the poultry fattener can buy this man's chickens at twelve or fourteen weeks old at 1s. 6d. or 2s. apiece to fatten, and can sell them again in three weeks at from 3s. 6d. to 4s. 6d. each, he also can make a profit.

As far as the farmer is concerned the question still remains, Can he find the necessary apparatus, labour, rent of land and food, and produce fowls at even 4s. apiece at a profit ? I very much doubt it, and I still further doubt if such prices as I have mentioned can commonly be obtained in the wholesale market. In egg-production I do not think the case is much better, for although the few eggs we get may be sold at 7 or 8 a shilling in December or January, they have to be sold at 20 a shilling for a longer period during the summer when our hens are laying well. If poultry fanciers, instead of devoting their time to the production of two white feathers in a particular place, and giving prizes for such nonsense, were to devote some time to the production of winter layers, the world at large would think better of them.

I have heard a few farmers say their poultry paid, but these have been men whose wives gave hours a day to looking after them. Turkeys, when properly managed, undoubtedly pay well; geese will get their own living on a field of rough grass, and may give a return for the grass eaten; ducks must be produced very early if there is to be any chance of them paying. It is my opinion that beyond keeping a few fowls in two or three movable roosts in the fields, the farmer will be well advised to let poultry-keeping alone. When running about the stackyards and feeding sheds they often do more damage to food than the whole lot are worth. I may be entirely wrong, but I cannot help feeling that the farmer has more serious matters to attend to, matters upon which the expenditure of the same amount of labour and attention would give greater results than is possible with poultry.

Poultry must be produced in this country by a similar class to that which produces it on the continent— the wife of the man who tills an allotment or small holding. Kept in numberless small batches where they can pick up the bulk of their own living, each batch may produce a small profit, but I have yet to be convinced that the farmer can keep poultry in sufficient numbers to pay for their accommodation and the time and trouble expended on them in anything like the same proportion as pigs, sheep, or calves.

HOUSE FED

CHAPTER XI
Will it Pay?

I HAVE endeavoured in this book to keep two facts
prominently in view: firstly, that the conditions of farming as a profitable industry have completely changed
during the last twenty-five years; and secondly, that
all the methods recommended must be such as are likely
to pay under the new conditions. We know to our cost
how great has been the change in the relative values
of corn and labour, and we cannot too carefully remember that our methods of farming, our tillage,
manuring, cropping, rotations, and stock-keeping, as
well as our legislation, customs, and methods of valua-
tion, have all been based upon the experience of a time
never likely to repeat itself.

That a considerable number of men make farming
pay at the present time is undoubted, but the successful
farmers are those whose business acumen, knowledge,
and skill have enabled them as far as possible to adjust
their practice to the change. I have already pointed
out that the most successful men always seemed to me
to be those who sold the most off their farms, and it
is worthy of special emphasis that these farms were
invariably clean, well tilled, and in good condition. This

is of course diametrically opposed to all the old ideas of farming, and there still exists a kind of righteous indignation against the man who sells other products than corn and a few fat beasts or sheep. The farmer must live and pay expenses out of what he sells, and yet most farm agreements and all valuations of tenant right are so framed as to discourage selling and to encourage buying, especially the buying of farm products from the foreigner.

The system of course originated in the days when wheat was worth from 60s. to 70s. a quarter, and as stock were merely kept as manure-producers, it would have been absurd to devote any of the valuable acres at home to their maintenance. At the present time it is quite different: corn is now so low in price that we may devote many acres to the keeping of stock, or the production of anything else that seems likely to pay; and as we now know of many other means of keeping up the fertility of the land besides the consumption of foreign foods, we do not need to grow our crops in strict rotations, which may have been useful when corn was the chief object of farming and dung the only fertiliser.

For farming to be profitable the farmer must be able to sell anything which it will pay him to produce. It may be urged that such a course would lead to the exhaustion of the soil; but this is not the case. In the first place, to grow profitable crops the land must be kept clean and well tilled; and secondly, the farmer is sure to find it pay him to use manures and to keep stock. I think I was able to show that the fertility of the soil depends far less upon the manure we add to it than on the way in which it is tilled, and there is no doubt that land is injured far more by incompetent

tillage and the unchecked growth of weeds than by the
selling of any reasonable quantity of hay, straw or other
produce. It is well known that it costs the farmer very
much more to keep his land clean than to manure it; and
yet, by our ridiculous customs, a farmer who took a foul
farm and thoroughly cleaned it would be entitled to no
compensation on that head, and would very likely be
mulcted in damages if he had sold a few tons of straw;
whereas the use of a few tons of cake would entitle the
slovenly farmer to a compensation which he in no way
deserved.

I am very strongly of opinion that present-day
farming will never be put upon a sound business
basis until it is freed from ancient customs and restric-
tions. Stock-keeping must be the mainstay of modern
farming, but the farmer cannot keep a large head of
stock all the year round while he is forced to grow
certain crops in certain fields in a definite order. To be
able to keep stock on the produce of the ploughed land
in the way I have indicated in the chapter on stock-
keeping, it is necessary to grow the crops required as
food for the cattle as close to the farmyard as possible,
while it may be necessary frequently to devote more
distant fields to the production of straw crops. Again,
in order to grow the largest possible crops and to obtain suitable forage crops exactly at the time they are
wanted, it is often necessary to use a quickly acting
manure like nitrate of soda, as well as other chemical
fertilisers.

There still exists a very strong prejudice against
chemical fertilisers, especially nitrate of soda, although
plants can take up their food from the soil in no
other form than that of a soluble salt, and their nitro-

gen only as a nitrate. As we know that even in
cake, that most sacred of manurial substances, all the
nitrogen has to be converted first into ammonia and
then into a nitrate before it is of any use to the plant,
the absurdity of this can be seen; but that R. M.
Garnier, in his " History of the English Landed Interest," should compare the use of nitrate of soda and
sulphate of ammonia to dram-drinking, shows how
strong this prejudice is. How unreasoning such a
prejudice may be is shown when the same author
deplores the waste caused by allowing the liquid products of the manure-heap to run down the ditches, a
point on which I thoroughly agree with him; but he
apparently forgets that practically the only manurial
substance in the liquid portion of manure is ammonia,
the use of which he has just compared to dram-
drinking.

The objection to growing two or more consecutive
crops of the same kind belongs to the same category.
In a state of nature plants commonly grow in the same
soil for generations, and by the decay of their remains
the top soil may become richer and the luxuriance of
the plants greater, owing to the increase in the supply
of readily available food within easy reach of the roots.
With our agricultural plants, if we constantly remove
the crop from the land, there comes a time when the
amount of food rendered available by natural causes
during the season ceases to be enough to grow a profitable crop; but it has been well proved that, if we
supply the necessary elements of plant food in such
a form as to become available in sufficient quantity,
land will grow the same crop for as many years as we
wish, without harm or detriment. Not only is this the

case when the soil food is supplied in the form of farm-yard manure, but it is also the case with properly balanced dressings of chemical fertilisers. This is well shown at Rothamsted, where wheat has been grown on the same land for fifty-nine years; indeed one of the chemically manured plots has produced a larger average yield during this period than that annually dressed with farmyard manure. In his " Investigations on Rothamsted Soils," Dr. Bernard Dyer has proved that during the twenty-eight years from 1865 to 1893, those wheat plots which have received complete dress-ings of chemical manures have not only become richer in phosphoric acid and potash, but the soil has lost very little of its natural store of nitrogen. The tendency seems rather to have been for the nitrogen to accumulate during the first few years, and in no case has the loss of nitrogen during twenty-eight years been more than would be supplied by a single applica-tion of about 15 tons of farmyard manure. From this we can see that there has been no great exhaustion of the soil under these conditions of manuring, although something like 4 quarters of wheat and 30 cwts. of straw per acre has been removed from these plots every year since 1843. I must not dwell longer on this subject, but any one who studies the matter will be able to see that no injury is caused to the soil either by the continuous growth of the same crop or by the proper use of chemical manures.

In practical farming, however, it certainly pays better as a rule to alternate our cereal crops with leguminous and other green crops, but we must beware of suppos-ing, with the agriculturists of fifty years ago, that different crops take different substances from the soil.

All crops take the same substances from the soil, although in somewhat varying proportions. Except with nitrogen in the case of the leguminous plants, all the green and root crops remove far more from the soil than the cereals, and it is only by consuming these crops on the land that their growth results in any benefit. That benefit is not due to any gain of plant food, but to the greater availability of the substances left in the top soil. We must, however, very distinctly remember that having by some means provided a supply of available plant food, it is our duty at once to grow a crop which shall utilise and lock up that food within itself, for nature will not allow any accumulation of soluble plant food in the soil. Investigations have shown that unless they are immediately used by plants, the lime, magnesia, potash, soda, chlorine, sulphuric acid, and nitrogen, as soon as they become soluble, are liable to be washed out of the soil by rain; in fact on heavily manured land more of these substances may be washed away and lost than are used by the crop. It is evident from this that after we have provided a supply of available plant food by cultivation and manuring, it is a true economy of the resources of the farm as far as possible to keep every acre constantly under a crop.

I have touched on this subject because I consider it essential to success that a farmer should be able to grow whatever crops he requires in the most convenient place and by the most economical method, and I have tried to show that this may be done without detriment to the farm. The farmer who always endeavours to grow the largest possible crops upon every acre of his holding is bound to increase the

fertility of his land.   He must keep his land clean,
he must cultivate thoroughly and manure frequently,
and if at the same time he can keep a fair head
of stock, the increase in fertility of that land must be
rapid.

The farmer who continuously grows large crops,
even if he sells some of them off the land, is bound to
have his farm in better condition than the man who is
content to produce small ones by what are considered
more orthodox methods.   The production of large and
fine crops contributes much to the pleasure of farming,
for there can be no satisfaction in contemplating a crop
which is a failure.   The farmer has also to look forward
to the realisation of his produce, and it is certain that
his pleasure will not be decreased by satisfactory profits.
What this produce shall be must depend very largely
on the situation of the farm and the personal inclina-
tions of the farmer ; but we have seen that there is no
good and sufficient reason why he should not dispose
of any particular product which offers him a profitable
return, provided he keeps his farm clean and uses the
means of producing good crops.   In certain districts
where wheat straw realises £3 a ton or hay £5 a ton,
no stock can be found that will pay to consume them
at such prices, and their manurial value can be returned
to the farm at a third of the price.   Near towns, the
sale of green crops such as lucerne, sainfoin, and tares
produce large amounts per acre ; potatoes are often
exceedingly profitable, and we may have an opportunity
of realising some of the profits from sugar beet, which
at present go to the continental farmer.   Corn growing,
I am afraid, can never again become a profitable
industry in this country, unless the straw and chaff

R

can be utilised to advantage. As an adjunct to stock-keeping it will still pay, but the crops must be large and the greatest economy exercised in their production. We know that during the last twenty-five years there has been a great reduction in the cost of corn growing, but there is still room for improvement. Harvesting machinery has resulted in a great saving, but there has been little corresponding improvement in the methods of tillage, especially in ploughing. The manuring is often far too expensive in proportion to its results, and more often insufficient through lack of live stock.

From whatever point we start, we seem bound to come back to the fact that stock-keeping is inevitably associated with profitable farming. This is well; for, were farming to realise the dream of the enthusiasts of fifty years ago, and to become simply an affair of steam ploughs and artificial manures, I am certain it would lose fully half its delights. The pleasure we feel in watching the growth of a crop can never be quite so great as that which we feel in patting a well-bred colt or calf which has ambled up to receive our caresses. We are apt to discourage intelligence in our farm animals: the cow that learns to open gates, or the sheep that creeps through the fences being quickly disposed of; but there is no doubt the Dutchman, whose cowhouse opens out of his sitting-room, or the Irishman, whose pig inhabits his cabin, both feel a pleasure in their animals much akin to that which we feel in the company of our dog. The farmer who takes a keen interest in his animals, not only has the pleasure of seeing them grow and thrive under his care, but the satisfaction of knowing that his profits are increased by his attention.

Dairying is an industry capable of realising considerable profits, particularly the selling of milk. In this there is as yet little fear of foreign competition, and the demand is constantly increasing. Cheese-making is worthy of greater attention, for high-class cheeses are both scarce and dear, but their manufacture requires a knowledge which can only be obtained by special training and practice. Butter-making is capable of great improvement in this country, but, owing to the enormous importation of foreign butter at low prices, the margin of profit is exceedingly small. It is only by our producing a first-class article and by the careful utilisation of the skim milk that any profit can be made; indeed it is doubtful even then if butter-making pays at all during winter. The greatest drawback to all forms of dairying is the difficulty of obtaining reliable milkers, and it still remains to be seen whether the milking machine is capable of providing a practical solution. Although the various branches of dairying are well worthy the consideration of all farmers, yet I am convinced that the full profits can only be realised by the man who is prepared to give its details his constant personal attention. It is for this reason that milk production is pre-eminently the industry of the small holder. If such a man can provide all the labour he requires from his own family, and devote his land to the production of food for his cows instead of trying to grow corn, there is no reason why he should not make as good a living off 40 or 50 acres as does his successful competitor in Denmark. It is, however, rather to the breeding and fattening of live-stock that the larger farmer will continue to give his attention;

more especially he who farms for pleasure. The great variety of interests introduced by this style of farming fills the whole year with pleasurable cares and excitements; and, the eggs not being all in one basket, some successes are assured. To supply the food and litter necessary for our horses, cattle, sheep, and pigs, requires the exercise of a good deal of skill and foresight in the arrangement of our crops. I have more than once pointed out the shortcomings of the four-course system, but it is difficult to recommend any particular course which shall be suitable to all cases. As a suggestion I will suppose that 100 acres of arable land have been cultivated on the four-course rotation, and a change is required to make it more suitable for stock-keeping. Under the old system, roots, barley, clover, and wheat would each occupy 25 acres. Suppose that 15 acres of the root area were given a small dressing of farmyard manure immediately after the removal of the previous crop, and after ploughing, 7 acres were drilled with winter beans and 4 with tares; and in the spring the other 4 acres were cultivated and drilled with mangolds. The remaining 10 acres of the root land, after a thorough cleaning in the spring, could then be manured largely with artificials, and sown with 6 acres of swedes, 2 acres of cabbages, and 2 acres of thousand-heads. The tares would be mown green or sheeped off by the end of June, the land thoroughly cleaned and perhaps sown with mustard, which, together with the beans and mangolds, would be cleared in time to sow 12½ acres of winter oats. The cabbages, swedes, and thousand-heads would be consumed during the winter and early spring, and the remaining 12½ acres

sown with barley. In the barley, red clover could be
sown, and amongst the oats a mixture of alsike,
white clover, and grasses. In the third year the
clover would be mown for hay and the second crop
sheeped off and afterwards sown with wheat, while
the grass mixture could be dressed with 2 cwts. of
superphosphate and 1 cwt. of nitrate of soda per
acre, and be either mown or partly mown and partly
grazed. In any case it would be grazed after mowing,
and with some roots and trough food, might be used
as a run for the ewes till February, when it would be
broken up and sown with oats. In the next course
the positions of the crops would be changed by
planting the beans and tares on the land formerly
occupied by the swedes, cabbages, and thousand-
heads. Numerous modifications could, of course, be
suggested according to the nature of the land, the
stock to be kept and the available pasture, but let
us see what such a rotation as this has given us. In
corn we have 25 acres of oats, 12½ acres of barley,
12½ acres of wheat, and 7 acres of beans, giving
us more corn and a far greater quantity of useful
fodder than under the old system. As a manured or
sheeped-off crop comes between every cereal crop
the yields of both corn and straw should be large,
and from 17 to 20 acres of hay should be sufficient
for our requirements. The root area is small, but if
the crops receive proper tillage and manuring, the
quantity grown per acre can easily be more than
double that usually obtained. The cabbages should
be ready for consumption in October and November,
the swedes in December, January, and February, the
mangolds in February, March, and April, the thousand-

heads in April, the tares in May, the grass mixture
in May and June, with the lattermath of the clover
and grass for July, August, and September. The
artificial manure required should not exceed 4 tons
of superphosphate per annum, with perhaps one or
one and a half tons of nitrate of soda, and if all the
beans and about one-fourth of the other corn were
consumed on the holding, it should rapidly increase
in productiveness and keep a large head of stock
without the purchase of any foodstuffs.

A farm cultivated on some such system as this, and
having about one-third of its area good permanent
pasture, is my idea of a model farm. Here we can
keep a few cows to breed and to supply milk for
weaning the extra calves we require. We can have
the pleasure of seeing these calves grow into fine
young cattle to be eventually fattened for the butcher
entirely on the produce of our farm. A breeding
flock can be kept, and the lambing of the ewes is
always a time of pleasurable excitement. The crops
are growing for their food, the lambing shelter has
been prepared, and we are expecting to be rewarded
for our care and forethought. At no other time of
year is there a stronger feeling of hopefulness than at
lambing; the increase in our flocks, the lengthening
days, and the knowledge that spring is approaching
fills us with a new feeling of buoyancy. Our walks
round the fields have lacked interest since the last
sown wheat came up, but now we watch for the signs
of growth, the fields are prepared for the sowing of
oats and barley, and every day adds to the numbers
of our flock. Winter is not over by a long way yet,
and the cattle are still in the yards, but with the

lambing begins that spring-time bustle which keeps
life full of interest the whole year through.

A farm like this, well-stocked and tilled in such a way
as to make every acre bring forth the utmost crop, will
require more capital and more labour per acre than the
larger area of half-neglected fields. It will, however,
bring in a greater return on the outlay than do the
large holdings one so often sees half-stocked and badly
tilled. Large farms are occasionally well tilled, and by
means of labour-saving machinery may be made to pay;
but they are neither so good for the landlord, the tenant,
nor the labourer, as the smaller farm highly cultivated,
nor are they so productive of wealth to the country at
large. The creation of small holdings has generally
been avoided owing to the expense entailed in the
erection of buildings, and this might very well be so
while the owner, architect, and builder hold the present
notions of what constitutes a set of farm buildings.
Occasionally I have seen a good set, but as a rule
needlessly expensive buildings are scattered about a
piece of ground in an apparently promiscuous manner.
Under the present changing conditions of agriculture
no landlord should erect expensive permanent buildings;
the chances are that in fifty years' time they will not be
at all suitable or even required. In default of a better,
let me make a suggestion as to premises for a small
holding. The whole set consists of a huge barn with a
concrete floor, wooden walls and corrugated-iron roof.
Taking 24 feet as an easy span for a roof, this barn
should be 48 feet wide, 60 or more feet long, and not
less than 15 feet high at the eaves, and should have large
doors at each end. A partition across the middle
divides this building into a stock-yard and a stack-yard

all under one roof. In the stock portion partitions about 6 feet high are run round at 12 feet from the walls, and with two or three doors this partition forms stable, cowhouse, calf pens and piggeries, while the space in the centre, 24 feet wide, forms a covered yard. The internal arrangements can be modified to any extent: a granary and mixing floor can be erected overhead, liquid manure and rain-water tanks provided; but in any case the stock, their manure, and their food are thoroughly protected, and the expense will be small in comparison with the accommodation provided.

There is a great demand for small holdings, and it is likely to increase as a better knowledge of their cultivation spreads. Farms of from 40 to 120 acres command high rents, increase the prosperity of a district, and in no small measure add to the well-being of the nation. Small or even medium-sized farms have not generally been very profitable during the last few years, but I believe that with improved methods of culture and greater attention to stock-keeping a new era will commence.

Given some prospect of success, there is no life so healthful and so full of pleasures as that of the farmer; but, to earn that success, business methods and a considerable amount of forethought, knowledge, and skill are necessary. In the old times it used to be the custom to make the fool of the family a farmer, but it would be a poor compliment to the most delightful of all occupations to suppose that such is now the case. Education is the necessity of the farmer; not necessarily a brain full of sciences; but the education which trains the mind to reason correctly and think ahead. The cloud which has so long hung over the British farmer will be a very long time moving away from him who is content to jog along the olden ruts. High prices for corn will never return; but I believe, nay I am sure, that the farmer who keeps well to the front in the improvement of his practice will find himself enjoying a certain amount of sunshine.

INDEX

THE END

Printed in Great Britain
by Amazon